DATE DUE

Unless Recalled Earlier

DEMCO 38-297

Value Analysis
in Design

Value Analysis
in Design

Theodore C. Fowler

VAN NOSTRAND REINHOLD
_____ New York

Copyright © 1990 by Van Nostrand Reinhold

Library of Congress Catalog Card Number 89-27638
ISBN 0-442-23710-3

Printed in the United States of America

Van Nostrand Reinhold
115 Fifth Avenue
New York, New York 10003

Van Nostrand Reinhold International Company Limited
11 New Fetter Lane
London EC4P 4EE, England

Van Nostrand Reinhold
480 La Trobe Street
Melbourne, Victoria 3000, Australia

Nelson Canada
1120 Birchmount Road
Scarborough, Ontario M1K 5G4, Canada

16 15 14 13 12 11 10 9 8 7 6 5 4 3 2 1

Library of Congress Cataloging-in-Publication Data

Fowler, Theodore C.
    Value analysis in design/by Theodore C. Fowler.
        p.   cm.
    ISBN 0-442-23710-3
    1. Design, Industrial.   2. Value analysis (Cost control)
I. Title.
TS171.4.F69   1990
745.2—dc20   89-27638
                                                      CIP

─── VNR COMPETITIVE MANUFACTURING SERIES ───

Product and Process Design

PRACTICAL EXPERIMENT DESIGN by William J. Diamond
VALUE ANALYSIS IN DESIGN by Theodore C. Fowler
A PRIMER ON THE TAGUCHI METHOD by Ranjit Roy
MANAGING NEW-PRODUCT DEVELOPMENT by Geoff Vincent
ART AND SCIENCE OF INVENTING by Gilbert Kivenson
RELIABILITY ENGINEERING IN SYSTEMS DESIGN AND OPERATION by
Balbir S. Dhillon
RELIABILITY AND MAINTAINABILITY MANAGEMENT by
Balbir S. Dhillon and Hans Reiche
APPLIED RELIABILITY by Paul A. Tobias and David C. Trindad

Manufacturing (hard)

INDUSTRIAL ROBOT HANDBOOK:
CASE HISTORIES OF EFFECTIVE ROBOT USE IN 70 INDUSTRIES by Richard K. Miller
ROBOTIC TECHNOLOGY: PRINCIPLES AND PRACTICE by Werner G. Holzbock
MACHINE VISION by Nello Zuech and Richard K. Miller
DESIGN OF AUTOMATIC MACHINERY by Kendrick W. Lentz, Jr.
TRANSDUCERS FOR AUTOMATION by Michael Hordeski
MICROPROCESSORS IN INDUSTRY by Michael Hordeski
DISTRIBUTED CONTROL SYSTEMS by Michael P. Lukas
BULK MATERIALS HANDLING HANDBOOK by Jacob Fruchtbaum
MICROCOMPUTER SOFTWARE FOR MECHANICAL ENGINEERS by Howard Falk

Manufacturing (soft)

WORKING TOWARDS JUST-IN-TIME by Anthony Dear
GROUP TECHNOLOGY: FOUNDATION FOR COMPETITIVE MANUFACTURING by
Charles S. Snead
FROM IDEA TO PROFIT: MANAGING ADVANCED MANUFACTURING TECHNOLOGY
by Jule A. Miller
COMPETITIVE MANUFACTURING by Stanley Miller
STRATEGIC PLANNING FOR THE INDUSTRIAL ENGINEERING FUNCTION by
Jack Byrd and L. Ted Moore

SUCCESSFUL COST REDUCTION PROGRAMS FOR ENGINEERS AND MANAGERS by
E. A. Criner
MATERIAL REQUIREMENTS OF MANUFACTURING by Donald P. Smolik
PRODUCTS LIABILITY by Warren Freedman
LABORATORY MANAGEMENT: PRINCIPLES & PRACTICE by
Homer Black, Ronald Hart, Orrin Peterson

Materials Management

TOTAL MATERIALS MANAGEMENT: THE FRONTIER FOR COST-CUTTING
IN THE 1990S by Eugene L. Magad and John Amos
MATERIALS HANDLING: PRINCIPLES AND PRACTICE by Theodore H. Allegri, Sr.
PRACTICAL STOCK AND INVENTORY TECHNIQUES THAT CUT COSTS AND
IMPROVE PROFITS by C. Louis Hohenstein

This book is lovingly dedicated to my wife, Dee, and to the six lights of our lives: Deb, Ken, Lin, Cin, Gin, and Barb.

CONTENTS

FOREWORD

Few books have been published on value analysis (VA) and value engineering (VE). If we consider those that emphasize the unique contributions of VA and VE, there are still fewer books, and many of these are out of print. Therefore, it is a pleasure to write the Foreword to this new text. As I have been mentioned and quoted here several times, it is appropriate to furnish a historical perspective of this subject and my relationship to the author.

I first met Ted Fowler in the late 1960s when he worked in the VA section at the Xerox Corporation. I had been invited to speak to the group about a new subject: customer-oriented value analysis. Initially, the group's understanding of accepted VA techniques and my concepts of customer approval clashed. Gradually, the group did understand the need for this new concept.

However, trying to sell the market research personnel at Xerox on the techniques used to determine customer attitudes was futile. Similar attempts to convince the product manager that competitive function analysis was a critical tool also failed. They repeatedly claimed, "We know what the customer wants better than the customer." Shortly after this attempt to introduce customer-oriented value analysis to Xerox, the company's VA section disbanded. I repeat these events because they illustrate U.S. industry's general response to VA and to customer-oriented VA.

Zealots are not easily discouraged, and Ted Fowler was a zealot. During long and emotional sessions, Ted and I worked to incorporate the customer-oriented functions into the Function Analysis System Technique (FAST). The results have stood the test of time. I still, however, prefer to call the primary supporting functions "satisfy customer" and "attract customers" rather than Ted's "enhance product" and "please senses."

The list of case histories in chapter 13 attests to the wide variety of products to which this approach has been successfully applied.

This book adds considerable depth to areas Muthiah Kasi and I only mentioned in our text *Function Analysis: The Stepping Stones to Good Value,* published in 1986. Our book concentrates on the techniques used in the information phase: Fast-diagramming, function cost, function attitudes, and function analysis. Fowler's new text builds on ours and fills in the important gaps by focusing on the remaining steps in the job plan.

Ted Fowler has spent many years experiencing the challenges of using the value analysis techniques. *Value Analysis in Design* succinctly weaves those experiences into a text from which there is much to learn.

THOMAS J. SNODGRASS
ASSISTANT PROFESSOR, UNIVERSITY OF WISCONSIN

PREFACE

The single objective of modern value analysis is to deliver to the user/customer the required functions at minimum cost.

The focus of the process is on harnessing teams of key employees to identify problems and then apply the appropriate problem-solving systems. It is thus an "umbrella" system, ensuring that the organization will most efficiently apply its resources of people, plant, and operating systems.

Various organizations have created such terms as *value management, value assurance,* and *value engineering* to describe the process for their own purposes. The correct term, and the one used throughout this book, is *value analysis.*

The value analysis process, developed at the General Electric Company, has a 35-year record of dramatic product improvement in thousands of organizations in the United States and 38 foreign countries. Modern value analysis, the subject of this book, reflects major improvements in that GE process. The focus shifts from product to customer, and the level of the problem solving shifts to the decision maker. The original emphasis was on *teaching* value analysis, in the hope that employees would apply the process to their everyday tasks. This, for many reasons, simply does not work. Today's emphasis is on solving specific problems. This results in participants who are taught very effectively and are also motivated to apply the process to the problems of their organization. For over fifteen years, several hundred organizations have applied this modern value analysis. It has invariably resulted in significant improvements in both user acceptance and product costs.

The goal of this book is to stimulate and guide U.S. industry toward adopting modern value analysis as an integrated element in the everyday corporate work process, with the ultimate objective of significantly improving the value of American products.

This book is written for engineers and engineering management by an engineer who entered the world of value analysis in 1955, at the peak of General Electric's love affair with its dramatically successful new "management tool."

The corporate euphoria lasted several years. In 1964, the "Father of Value Analysis," Lawrence D. Miles, retired from GE and moved out into the world with his young "child." From that date, the value analysis system has constantly fought a generally successful battle for survival, initially within GE and later throughout the world. Within GE, though, value analysis today is not a significant program. In the spring of 1964, in fact, Miles's replacement, William Kuyper, overstated to this writer, "Value analysis is dead at the General Electric Company."

Many of the value analysis practitioners who had moved out from GE into the wider world of U.S. manufacturing refused to accept this obituary notice, and have since built a record of success, expanding the techniques for application in nonmanufacturing, construction, government, and, more recently, most of the world's other industrialized nations.

The record of value analysis in optimizing the value of existing products is unexcelled. Its power is even greater, though less publicized, where it is focused on products in the early concept stage, before any expenditure for planning or tooling. These so-called upstream successes have resulted in the development of a sophisticated system for establishing and analyzing a product base case method to simplify comparison and measurement. (A base case is a rigorous set of drawings and a detailed, costed bill-of-materials. It assures a firm, nonmoving base reference for the analysis.)

Nevertheless, the rumblings have continued. Hundreds of top managers throughout the world have echoed Bill Kuyper's pronouncement. While many value analysis systems have thrived and returned outstanding benefits to their organizations, the majority of the value analysis systems established in the United States have failed, usually within the first 18 months.

As a professional, committed, and generally successful value analyst, this writer found the wholesale failure rate disturbing. By the late 1970s, with the backing of the Xerox Corporation and later the enlightened support of Thomas J. Snodgrass and other committed value analysis people throughout the country, he resolved to put together a modern value analysis system, which would modify and expand the Miles system and eliminate most of the reasons value analysis systems fail.

This book describes the new system that has evolved. It is basically "Miles-ian." It recognizes, however, that the original value analysis system contained many of the seeds of its own destruction. These have generally been weeded out. It also takes significant advantage of the 30 years of change in the business world since the original value analysis system worked its wonders.

Unfortunately, much of the value analysis used today follows the original GE model. For this reason, there remains much disillusionment scattered around the country and the world. To fail to acknowledge this rocky environment would be to ignore the hard-nosed realities regularly faced by both the value analysis practitioner and those who are instituting their own value analysis systems.

Throughout this book, there are comparisons between modern value analysis and the original value analysis on which it is based. Miles has said that "if there is no comparison, there is no evaluation" (Miles 1961). These comparisons should not be judged as complaints or accusations, but rather viewed as placing in perspective the rigor and the vigor of the new system.

ACKNOWLEDGMENTS

Each case study in this book is a direct and valid copy of an actual Value Analysis study, or an in-context adaptation of such a study; led by one of the following value specialists: Theodore C. Fowler, Thomas F. Cook, J. C. Boyers, Robert Null, Ronald Rathsam, Fred D. Hauke. Appreciation is expressed for the data provided to the author for his selection and adaptation for emphasis.

Chapter 1

THE DEVELOPMENT OF VALUE ANALYSIS

THE ORIGIN OF VALUE ANALYSIS

Value analysis is a system for harnessing people first to identify problems and then to select and apply the appropriate problem-solving procedures.

It was born in the late 1940s at the General Electric Company. It has evolved in the succeeding 40 years or so from a purchasing-based analysis method into a management-based system for optimizing the worth and cost of various functions.

In 1947, Lawrence D. Miles (Fig. 1–1), a GE electrical engineer in Baltimore, requested reassignment to corporate purchasing in Schenectady, New York, to develop his 'function-oriented" concepts into an orderly system. His boss, William Shredenschek, assistant to the vice president of corporate purchasing, supported Miles's efforts and guided the development of the new system, now named value analysis.

Its impressive successes at Schenectady and in the 92 GE departments throughout the United States resulted in the training of thousands of employees in the basics of value analysis. Three million dollars was invested by 1958 in the development and extension of the technique. When the GE program was deemphasized in 1964 with Miles's retirement, the corporation judged that the new value consciousness of thousands of participants was a proper return on this investment.

In the mid-1950s, several major U.S. companies set up their own value analysis systems, followed by the Army through the Watervliet Arsenal in New York and the Navy through its Bureau of Ships.

Japan picked up value analysis with enthusiasm in the early 1960s. As in other areas, it has beaten us at our own game. Essentially every company there now has a value analysis system.

Figure 1–1. At the Air Force Museum, Lawrence D. Miles views a B-29 turbo-supercharger, the item he was required to buy for General Electric in 1942, setting the basis for his discovery of function analysis.

Value Analysis Expands and Changes

In federal government usage the process was renamed value engineering, for the eminently practical reason that it was easier and more rewarding to staff the effort with engineers than with analysts. Miles credits Admiral Wilson D. Leggett (Fig. 1–2) with the creation of the new name. Impressed with the results of a value analysis seminar in Washington, D.C., with Roy Fountain, Miles's chief collaborator, Leggett decided the Navy would initiate a value system, but he told Miles he would have to call it value engineering in order to ease it in at a sufficiently impressive level.

This change in name coincided with the development of a professional society of value analysts, named the Society of American Value Engineers (with the highly appropriate acronym SAVE). With this, in many quarters, value analysis became known as value engineering.

A substantial body of opinion today holds that this name change compromised the growth of value analysis by implying that it is an engineering discipline, rather than a universal system guiding people to identify problems and then select and apply the appropriate means of solving them. Whether the new name has been good or bad, there is no question that change has spawned more change, until now there are no fewer than a dozen different names for the same process.

Figure 1–2. Admiral Wilson D. Leggett, who gave Miles's value analysis system the alternative name value engineering.

What's in a Name?

The subject of this book is a process called value analysis. Then what is value engineering? Or value management? Or value control? Or the dozen or more other names for systems that appear to be describing the same process?

When the U.S. government created the term *value engineering*, it set the precedent for an explosion of names, including not only the ones mentioned above, but also value assurance, value improvement, VA/VE, value marketing, value standards, value research, value studies, and value purchasing.

All of these are in current use. For various reasons of varying validity, each is judged by its users to be more useful to their purposes. Whatever the name, the process used in each of them is still value analysis.

Don't Just Change the Name on the Door

In some quarters, value analysis is regarded as no more than a modern opportunistic catchphrase. Frederick Sherwin, an early GE and Raytheon value analyst, saw many cases in which an industrial engineering department was renamed a value analysis department. His constant refrain in presentations to management was "Don't just change the name on the door!" "Value analysis is unique," said Sherwin. "So is industrial engineering. Misidentifying either of them weakens them both."

The Government Value System

Most value analysis has been performed in the private sector of the economy. The U.S. government's value engineering system, like many other government programs, tends to rise and fall from time to time. The Public Buildings Service of the General Services Administration developed a strong program in the 1970s. More recently, the Department of Defense has demonstrated an increase in activity. Congress may soon consider a bill that would require the use of value engineering techniques by government agencies and many organizations doing business with the government. A recurring problem in government value systems, though, is a lack of focused motivation. Most government managers respond to the primary pressures of performance and the budget, which in turn are influenced by legislative and executive pressures. There is seldom a direct motivation to increase product value.

A second problem is that the pertinent government regulations were written to permit nearly any cost-reduction effort to be called value engineering. This dilution of the term has misidentified the value analysis process in many quarters.

Construction Value Analysis

The original government regulations authorizing the inclusion of value engineering in Defense Department contracts contained a restriction in nearly every paragraph: "except construction." This was removed in the early 1960s, chiefly through the efforts of a Navy professional engineer, now a prominent value engineering consultant, Alphonse J. Dell'Isola.

In 1971, Donald E. Parker of the General Services Administration established a program to reward contractors for value engineering improvements in government construction. The program expanded in 1973 to include government funding for a value engineering effort. This has spawned a major expansion of value engineering throughout the architect-engineer and construction management community that continues today.

Both the construction and, to a lesser degree, government value engineering systems in general suffer from the second-guessing syndrome. Most of these value engineering efforts are performed by outsiders; that is, in terms of modern value analysis, the key man is not included on the value team. As a result, the team's conclusions often must be implemented in the face of opposition by cognizant engineers. This, in turn, results in much less than 100 percent implementation of recommendations and, more important, tends to build an active antipathy toward the program. But construction value engineering has risen above much of this opposition. Indeed, the so-called Midwest school of construction value analysis, led by Thomas J. Snodgrass, Muthiah Kasi, and Howard Ellegant, applies the most modern form of user-oriented value analysis and regularly accomplishes remarkable results.

An example of the power of modern value analysis in construction is Marillac House in Chicago (Fig. 1–3). Howard Ellegant, AIA, CVS, led a panel of 12 nuns, three lay persons, and the architect in establishing the functions or performance requirements for the structure. In a creative session with the architect, a building then was defined that fulfilled all user needs at an affordable cost. Ellegant, a particularly enthusiastic supporter of modern value analysis, often approaches a client with a question: "Do you want excellent proposals—or do you want excellent proposals that will be implemented?"

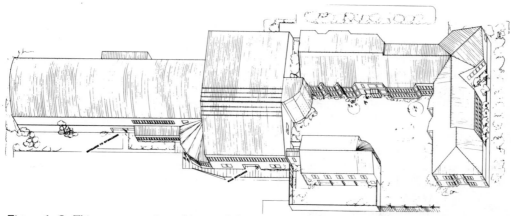

Figure 1–3. *This axonometric architectural drawing shows the new Marillac House community service facility of the Daughters of Charity, in Chicago; the building resulted from a value engineering study.*

His thesis is, quite simply, that the common second-guessing form of value engineering in construction is often an exercise in unimplemented futility.

Private-Sector Value Analysis Systems

The expansion of value analysis throughout private industry since the 1950s has taken a very different course. Here the motivation is much clearer: profit. Management supports the value analysis system because it promises to contribute directly to its prime objective: the return of a profit on an investment.

Driven by this clear motivation, hundreds, then thousands, of companies, schools, hospitals, and service organizations established value analysis systems.

Unfortunately, although the promise is great, and the initial results are often dramatic, the great majority of these systems have failed, and private industry can boast a record of overall accomplishment only marginally better than that of the government. There are presently fewer than 10,000 organizations with active value analysis systems in the United States.

The reasons for the many failures are varied, but they tend to result from lack of recognition of an important fact: the value analysis system is an integrated set of procedures that requires intelligent support and careful nurturing. A heavy-handed or inexperienced approach will destroy the best-planned value analysis system.

A detailed description of 20 principles that guarantee a successful value analysis system is provided in the Epilogue to this book. Failure to comply with a significant number of these principles can seriously inhibit the system. With intelligent planning and attention to detail, however, value analysis invariably realizes its true potential.

Value Analysis in Japan and Beyond

In the early 1960s, the Japanese picked up value analysis through their association with U.S. companies. As early as 1963, the definitive value analysis book, Miles's *Techniques of Value Analysis and Engineering,* was translated into Japanese. In 1967, a value engineering project workbook, based on a Raytheon Company publication, was published in Japanese. The process has become a significant force throughout Japanese industry. The Society of Japanese Value Engineers accepts only companies as members and serves as the central agency for education and promotion. There are presently 9,000 practicing value analysts in Japan. In the larger companies, it is common for the top operating executive to perform as the company's value engineer.

In 1983, the Miles Award was established in Japan and presented to three of that nation's industrial giants for their work in cutting costs without sacrificing quality. In 1985, the Japanese government posthumously awarded Miles, who died that year, a High Order of Imperial Merit Medal in recognition of "his outstanding contribution to Japanese industry and economy in his life and through his VA/VE methods." This medal has been awarded to only three other Americans: Lillian Gilbreth, Peter Drucker, and W. Edwards Deming.

Other countries were also quick to pick up this dramatic new industrial problem-solving technique. There are now active value analysis efforts on all continents and in several dozen countries.

THE ORIGINAL VALUE ANALYSIS SYSTEM

Lawrence D. Miles developed value analysis through a deductive process:

First, he prepared a list of reasons for the unnecessary costs that he found in all products and processes:

- engineers' other responsibilities
- lack of time
- habits and attitudes
- lack of information
- preconceived ideas
- prejudice
- temporary circumstances
- lack of ideas
- lack of experience
- failure to use available specialists
- desire to conform
- fear of personal loss

Starting from this base, Miles and his GE value services group built a new philosophy. Some of its elements were old:

The four kinds of value, defined in 1929 by Correa Walsh

The value analysis job plan, based on the classic scientific method of Anaxagoras

The supplier workshop, an established purchasing technique

The application of good human relations, as defined by the new breed of sociologists

Dick Borden's Effective Presentation technique as heavily supported by GE

Brainstorming, being developed concurrently by Alex Osborn

Some elements were unique combinations of then-current knowledge, including the team workshop seminar of 40 to 160 hours, combining the interaction of a cross-disciplinary team with the classic participatory seminar and a "live" project whose results could be implemented to amortize the cost of training.

Some of the elements were new, such as the system of function analysis. This concept grew from Miles's initial experiences in purchasing. He found that if an item to be purchased were specified in terms of its functions rather than by a mechanical drawing, a better price resulted. Indeed, a better product usually resulted as well.

Miles developed this concept into the magic of his functional approach, with its focus on verb-noun function definition.

A second new concept was evaluation by comparison, in which Miles states that the value of anything is variable, depending on environment and attitude. Value can be quantified, at any instant, only by comparing it with another item performing that same function.

From these two new elements grew the original concept of function value, which states simply that the value of a product or process is the lowest cost that will provide its functions.

The Cross-Functional Team Workshop

When value analysis proved its worth in the early 1950s, General Electric engineering management instructed Miles to "train 1,000 people a year." This placed an early, and in many ways unfortunate, emphasis on training, which continues even today in many value systems. Miles and Roy Fountain set up a 160-hour seminar training system, later refined to 40 hours. This has served as a model for the training of nearly one million people throughout the world. This early seminar approach combined several features that multiplied the effectiveness of the value system:

Teaching: The teaching seminar has a successful history in academia as an interactive session focusing on a single subject.

The cross-functional viewpoint: Participants are selected from different areas of expertise to ensure that all viewpoints are included in the discussion.

Time-phased: The effort is organized into time-sequenced phases.

Live project: Real problems are chosen for analysis.

Self-funding: The implemented savings resulting from a value analysis workshop training session commonly exceed the cost of the effort.

Experience with thousands of value analysis workshop seminars has taught us that certain of these elements are essential to effective value analysis. But one element is counterproductive: the teaching focus.

The initial GE objective was to train large groups of employees to become "value conscious," hoping that they would then apply the techniques to their daily work. For several reasons, this carryover simply does not occur.

MODERN VALUE ANALYSIS SYSTEMS

In the 25 years since Miles first put together this remarkably effective combination, there have been numerous improvements. Many of these came from Miles himself, still the giant of value analysis at his death at age 81 in 1985.

A fundamental leap ahead took place at the Xerox Corporation in 1968 through 1970. This leader in technology invested and experimented without constraint in the application of new process tools and created the basis for what is now called modern value analysis.

Many other improvements have come from students and practitioners of value analysis, particularly from two fundamental contributors, Thomas J. Snodgrass and Charles W. Bytheway.

Snodgrass (Fig. 1–4), a General Electric Hotpoint engineering manager, proposed a shift in the customary view of value. The original Miles-ian concept that "only basic function has value" was, Snodgrass felt, difficult to reconcile with market realities. In

Figure 1–4. Thomas J. Snodgrass, now a professor at the University of Wisconsin, is the creator of user-oriented value analysis.

the late 1950s, he created an organization called Value Standards, Inc., through which he has since promoted his concept that "the value of each of the many functions performed by a product or process is determined by the user/customer's acceptance of each of those functions."

In 1963, Charles W. Bytheway (Fig. 1–5), a Univac value analyst/logician, developed a system that he called basic function determination technique. It contained a new function logic method, which he called function analysis system technique (or FAST), that dramatically improves the effectiveness of Miles-ian function analysis.

Other, more recent improvements making up modern value analysis include the following:

1. Value analysis projects are no longer identified by management in advance of the value analysis workshop. Management still identifies the specific product to be value-analyzed. The projects, however, are identified by the value analysis team, using techniques that pinpoint where the product does not match the users' needs and wants. The team then focuses its problem-solving efforts on those pinpointed value analysis targets, which Thomas F. Cook calls value mismatches.

2. Preparation for implementation is now shifted to the first step in the problem-solving process, rather than being merely an essential afterthought.

Figure 1–5. Charles W. Bytheway is the inventor of function analysis system technique, the system called FAST.

3. The previously standard 40-hour, one-week training workshop that GE developed in 1952 is now replaced by a 55-hour, eight-week schedule, which focuses on results.

4. The objective of the workshop is problem-solving. The training aspect is accomplished indirectly—and much more effectively—through the inevitable overwhelming success that the participants experience.

5. The 18 hours of stimulating lecture that characterized early value analysis training workshops, and still prevails in many, is reduced to little more than a 2-hour interactive case study plus periodic introductions to each phase of the job plan.

6. The focus of the workshop is shifted to the 40-plus hours of team interaction, closely guided by a value specialist. An additional 60 hours of outside effort reinforces the problem-solving focus of modern value analysis.

7. Rather than simply delivering their proposals to be implemented, the team members now accept ownership of these ideas by volunteering to champion the proposals, ensuring that each has the maximum opportunity for implementation. The author first encountered the champion principle in value analysis when Thomas F. Cook proposed it as an incidental element in his effective packaging of the value analysis system. It is now regarded as being of highest importance to the system.

Modern value analysis now precisely matches the "quick-hit, small-team" process that Tom Peters and Nancy Austin described in *A Passion for Excellence* as a powerful alternative to the "throw resources at it" approach to problem solving (Peters and Austin 1985, 195).

The Objective of Modern Value Analysis

The single objective is to deliver to the user/customer the required functions at minimum cost.

Objectives of the Modern Value Analysis Workshop Study

Major restructuring of the original General Electric teaching seminar has resulted from 25 years of effort by Thomas Snodgrass, Charles Bytheway, and the author.

The result is a modern value analysis workshop with four goals, in order of importance:

1. To improve user acceptance of the product or process under study.

2. To reduce the cost of the product or process.

3. To train the team members by having them experience an impressive success with their value analysis projects.

4. To establish value analysis as an ongoing system that ultimately will be formally applied to all problems of the organization that concern cost or function.

Concentrating on the first two goals, by focusing on optimizing the results of the workshop project, closes a learning loop and inevitably results in achieving the other two goals.

For this reason, the workshop is purely results-oriented. The rigor and the detail are all carefully calculated, interrelated, and balanced toward one objective: The workshop project must succeed. Cost must be minimized. User acceptance must be maximized. In short, value must be optimized.

This success closes the learning loop in the most effective way possible: by example. In a well-run workshop, the team members invariably reach a point of revelation. There is a special moment when they realize that, through their own efforts, in just a few days of concentrated effort, and by simply following the rules and the sequence of the value analysis job plan, they have dramatically improved the value of their assigned product.

Repeated successes build upon themselves, resulting in a continually broadening field of application of value analysis in the organization.

This transformation of an organization's method of problem-solving is gradual. Each participant enters a workshop equipped with a personal set of problem-solving procedures. It is unlikely that the team member will easily discard these methods, which were built up over a lifetime, during the brief period of a workshop, even when it is crowned with dramatic success. A workshop success, however, will plant the seed. With nourishment—including repeated successes—value analysis will become a way of life for employee and organization alike.

An organization that has methodically built value analysis into its fabric is the Butler Manufacturing Company of Kansas City, Missouri. The value analysis system at this multidivisional leader in pre-engineered buildings has focused the process on more than 25 products and procedures (Fig. 1–6). Each success has built upon the last, firmly establishing value analysis as the standard cost and function problem-solving system at Butler. The major key to its success has been the active, informed involvement of its president and the chairman of the board.

HOW TO ESTABLISH VALUE ANALYSIS

Avoid a "Standing" Value Analysis Team

It is not appropriate to establish a permanent team of value analysis experts. In the past, such a formal group has occasionally been established and assigned to value-analyze projects as required. This approach fails, simply because such a group invariably becomes involved in second-guessing other experts. While its proposals for change may be implemented on their merit, such an activity builds within the organization an active antipathy, which inevitably destroys the value analysis effort.

Product or Process

Value analysis was created to analyze hardware products. Its success has led practitioners to apply the method to procedures and processes as well. The results have been equally impressive. Indeed, value analysis is effective whenever it deals with anything that performs a function and costs money.

Because this book focuses on the concerns of the design engineer, the majority of the examples of value analysis involve products, not processes.

Figure 1–6. Value analysis at the Butler Manufacturing Company reduced the cost of its roof skylight/ventilator by 11.7 percent.

Please read the word *product* in the broadest sense. A process or a procedure is, in fact, also a product of the organization.

Quality—Today's Hot Button

Throughout history, people have tended to pattern their actions after others'. At some point in the history of Rome, one pattern was the wearing of togas. More recently, in U.S. industry, the major pattern was automation. Today, the industrial pattern world-wide is quality.

The emergence of Japan as a global industrial powerhouse beginning in the 1950s owes a debt to two Americans, Dr. W. Edwards Deming and Dr. J. M. Juran, and their message of statistical quality control (SQC). The results of this message returned to the United States in the 1960s and ever since, in the form of a flood of goods of exception-ally high quality. U.S. industries, particularly in the automotive sector, were stunned by this unexpected attack. After a time, many of them decided to join rather than fight, and the great American infatuation with quality was under way. SQC has been legitimized. It is no longer simply a tool of bespectacled mathematicians plotting least-squares curves in quality control laboratories. It has become the way to reassert our preeminence in the world industrial market.

Quality circles is a management system that descended upon us from Japan by way of Lockheed. Total quality control, as defined by Kaoru Ishikawa, Secretary General of the Union of Japanese Scientists and Engineers (JUSE), later became a hot subject in the boardroom. Total quality management is a variant that the U.S. Department of Defense recently embraced. The all-controlling complexity of quality function deployment is the modern way to structure an organization for new product development. Autonomation, developed by Shigeo Shingo at Toyota, became automated quality, or *poka yoke* (mistake-proofing). It appeared as if success in American industry required embracing at least one of the seemingly endless series of new management systems, all with the word *quality* in their title.

The message was not lost on value analysis practitioners. A common subject of discussion since 1981 at gatherings of value analysts has been: "How can we change our name or change our approach to include the concept of quality?" Papers were presented at national conferences of the Society of American Value Engineers and the Miles Value Foundation suggesting that the value movement would benefit greatly by including itself in the expanding quality movement.

A typical concern, raised repeatedly at SAVE conferences, takes this general form: Value analysis attacks the most fundamental relationship in modern history: that between worth and cost. Value analysis is presently used by about 10,000 organizations, approximately 5 percent of the total. However, quality programs attack only one small element of that relationship. Quality programs are in effect in nearly 100 percent of U.S. organizations. Therefore, value analysis should lock into this success by merging itself into the quality movement.

There have been suggestions that value analysis be attached to today's hot button by renaming the process or changing the worth-over-cost equation to something including the word *quality*. Each such suggestion fades quickly when the discussion progresses to the fundamental questions: "What is quality?" and "What is value?"

Judging from common responses, it would appear that neither question has received an unequivocal answer; but a review of some prime sources of word definition (Table 1–1) reveals a remarkable similarity in concept between the words *quality* and *value*.

The dictionary definition of quality presents a major problem in measurability. Philip B. Crosby, former ITT Vice President, drastically narrowed the definition and achieved measurability. The corporate definition drifts back into the immeasurable area, but retains the narrowness of the Crosby definition.

The three definitions of value in Table 1–1 appear to suffer from an even more serious immeasurability, but all are clearly broader in concept than the definitions of quality, including all of the "excellence" and "conformance" and "meeting of customer expectations" components. Further, the value analysis described in this book provides techniques to define the fuzzy areas of user needs and wants, removing the measurability problem.

A properly defined modern value analysis effort is a study of user perception of all the relationships of a product, including performance, cost—and quality. Value analysis

TABLE 1-1. COMPARISON OF DEFINITIONS OF QUALITY AND VALUE

Quality		Value	
Source	Definition	Source	Definition
Webster's Ninth New Collegiate Dictionary	The dictionary contains 14 definitions plus three synonyms for quality. The one that applies most closely to our concerns is "degree of excellence."	*Webster's Ninth New Collegiate Dictionary*	"The monetary worth of something." (Worth = "The value of something measured by its qualities or by the esteem in which it is held.")
Crosby 1979, 15	"Conformance to requirements"	Miles 1971, 4	"Appropriate performance and cost"
General Motors Corp.	"Conformance to requirements that meet customer expectations"	This book	Performance plus Quality (Supporting functions)
			First cost + Follow-on cost

assesses the quality of a product in terms of the four supporting functions which are found in all products: *assure dependability, assure convenience, enhance product,* and *please senses.*

In the words of Marquis de Vauvenargues, "Patience is the art of hoping" (Stevenson 1967, 1462). We are patient with those who would shift the objectives and the structure of value analysis to make it more marketable in today's environment. Each of them will eventually come to realize that value analysis is a system for harnessing people first to identify problems and then to select and apply the appropriate problem-solving procedures.

According to this definition, the process of value analysis is primarily behavioral. First it welds together a team. Then the team abstracts its viewpoint of the study product through function analysis. Next it identifies problems in the broadest conceivable human and industrial terms, those relating cost and user acceptance. Finally, it applies whichever available problem-solving systems are most appropriate.

Value analysis, therefore, is the umbrella system that makes optimal use of the nearly infinite number of formalized, focused problem-solving systems, including total quality control, quality function deployment, quality circles, autonomation, industrial engineering, creative problem solving, work simplification, Synectics, financial analysis, industrial design, Kepner-Tregoe, operations research, purchasing analysis, mathematical programming, critical path, econometrics, test marketing, forecasting, design for manufacturability, management-by-wandering-around—in short, any focused problem-solving system that is appropriate to the problem identified by the value analysis system.

Value analysis does not compete with focused problem-solving systems; it simply optimizes their application.

Early-Stage Value Analysis

Value analysis was developed within a purchasing environment. The method was conceived and proven through the analysis of hardware items that were then being manufactured. It was clear, even to early practitioners, that a much larger payoff lay in value-analyzing products before release to manufacturing or, ideally, even before the design concept is clearly established.

Figure 1–7 shows the typical life cycle of a manufactured product. The curves are arbitrarily scaled, but they clearly convey a fundamental business truth: The earlier in the design cycle that value analysis is applied, the larger the savings and the lower the implementation cost for any resulting change.

The obvious difficulty with performing value analysis upstream—in the concept, development, or design stage—is the problem of measuring effectiveness against an undefined or moving target. In order to prove that value analysis has succeeded, it is essential that there be a measurable *before* and *after*. It would therefore seem that, while upstream value analysis is desirable, it should be avoided because its effectiveness can never be proven.

But this is not the case. A system called the base case method has been developed to permit effective upstream value analysis. The method establishes a fixed target, in terms of both function and cost, even with products that are no more than a gleam in the eye of the marketer or engineer. It is described in detail in chapter 4.

The Policy Document

Although a value analysis system can succeed for a while in an organization on a pro tem basis, a written policy is essential if the system is to survive.

Appendix A contains a typical policy document for a multidivisional company. It establishes a total corporate value analysis and cost-improvement system, including the elements of value analysis, traditional cost improvement, scrap reduction, delivery cost reduction, and operations or methods improvement. An organization must create a similar policy document if the value system is to become an effective element in its day-to-day business practices.

Five structural elements must be included in any such document:

- a value council

- a reporting system

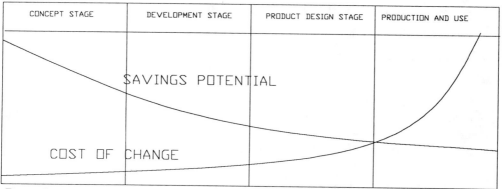

Figure 1–7. The two arbitrarily scaled curves indicate greater savings potential and lower costs when changes are made earlier in the product life cycle, before design and production.

- a follow-up system

- a monthly summary report

- an annual business plan system

An essential early step in establishing a value analysis system is submission of a copy of the policy document to the comptroller or vice president of finance, with a request that the executive tailor the document to the realities of the organization. The policy document is then distributed throughout the organization.

The Value Analysis Coordinator

The selection of a coordinator to lead the value analysis system is a critical step toward its establishment in the organization.

Among the criteria for selection are 1) a knowledge of modern value analysis, including participation in several successful value analysis workshop studies; 2) some experience in guiding teams in those studies; 3) a strong emotional drive to establish an effective value analysis system; and 4) a personality that stimulates interpersonal activity and creative problem solving in others.

The Value Analysis Council

Under no conditions should personnel be added or anyone's responsibilities measurably expanded in creating a value analysis council.

The council is merely a point of focus for the value analysis and cost-improvement system. It should comprise the top operating manager and key members of his or her staff.

Council meetings take the form of agenda items at the manager's regular staff meetings, during which the value analysis coordinator reports and requests guidance (Fig. 1–8).

The responsibilities of the council are limited to those shown below:

- identify products for value analysis

- authorize participants for value analysis workshops

- attend all workshop final presentations

- approve change proposal implementation plans

Reports

The focus of a value analysis reporting system must be on implementing proposals for change. Each of the forms included in Appendix A is carefully crafted toward that end.

The desired forms and their functions are described in Table 1–2.

Where Should Value Analysis Report?

Value analysis has successfully reported to nearly every functional area. While it is commonly regarded as an engineering activity, it has also succeeded in manufacturing,

Figure 1–8. *The value analysis coordinator uses a flip chart while briefing executives during a value council meeting in the top executive's office.*

TABLE 1–2. KEY VALUE ANALYSIS SYSTEM FORMS

Form	Function	Frequency
Value analysis proposal	Submitted by person proposing a change affecting cost	As needed
Value analysis and cost-improvement project list	Continuously updated status of each active proposal	Monthly
Value analysis and cost-improvement project summary	Report progress against plan by project	Monthly
Value analysis and cost-improvement system performance summary	Report progress against plan for system	Monthly
Consolidated system performance summary	Report to top executive on corporate status against plan	Monthly

quality assurance, and human relations. It has even succeeded for a time while reporting to the top executive.

The unique structure and the personality of an organization are the chief determinants of where value analysis should report. The choice should be based on four considerations:

1. When value analysis reports at an excessively high level in the organization, it implies authority. This implication tends to crumble the first time a request for

high-level backup collides with reality. The top-level manager cannot be concerned with decisions at the value analysis operating level.

2. Value analysis always affects design. In many organizations, this requires that the value analysis function have a seat of power in engineering. In an organization in which design responsibility has been honestly delegated to manufacturing, a production-based value analysis system may work well.

3. If the quality process is being or has been adopted as the way of doing business in a corporation, then the value analysis coordinator must make sure that value analysis is an integral part of that quality process. In this scenario, the value analysis function could report to the corporate quality office.

4. The finance group must be deeply involved in value analysis. Where the culture of the organization permits, this is probably the best home for the value analysis function. At the very least, finance must have total auditing authority.

The basic rule is to place the value analysis activity in the functional area that results in a minimum of provincialism, and at a sufficiently high level to assure appropriate independence and authority.

Chapter **2**

WHAT IS VALUE?

THE DEFINITION OF VALUE

A product must fulfill a user's need or want in order to have value. The most fundamental concept of modern value analysis, this is represented by the following basic equation:

$$\text{value} = \frac{\text{worth}}{\text{cost}}$$

This equation is totally valid. Further, it has the ring of good philosophy. But look a bit closer. How can this formula be used to measure value? Cost can be quantified in dollars, but what are the units of worth? How do you actually measure the level of worth in a product or service?

Value analysis solves this problem by first translating the product or process into the functions it performs and then measuring the user acceptance of those functions. This user acceptance is a direct measure of worth.

The equation is actually a bit more complex than simply worth divided by cost. The user perception of worth is composed of first impressions plus experience. Cost is not simply the initial price, but must also include follow-on costs during the life cycle of the product. The equation for modern value analysis includes these factors:

$$\text{value} = \frac{\text{user's initial impression} + \text{satisfaction in use}}{\text{first cost} + \text{follow-on costs}}$$
$$\text{(also referred to as total cost or life cycle cost)}$$

User perceived worth is a complex concept. Even the user alone is a complex element, not consisting solely of the final buyer or customer. Examples of simple products with several levels of users are listed below:

Product	User/Customer
	Car owner
Automobile headlight	Repairman
	Car manufacturer
	Wholesaler
Flashlight battery	Retailer
	Flashlight owner
	Policy holder
Insurance claim	Insurance company
	Victim

In addition to the complexities of trying to measure the acceptance of several different users, we are faced with the subjectivity of any user reaction (Fig. 2–1). That is, each individual user reacts to a product in keeping with his or her particular set of attitude screens. Further, these screens are constantly changing. With this complexity and variability, how is it possible to measure the user acceptance of a function?

Simply stated, this can be done by gathering a statistically valid set of responses from

Figure 2–1. Products have several levels of users, each with a particular set of attitude screens: here, not just the purchaser of flashlight batteries, but also the seller, who constitutes a second level of buying influence.

actual users who represent the prime buying influence; that is, the person or people whose viewpoints most affect product acceptance. Those responses are then allocated to the functions to which they relate.

The process is covered in chapter 6 under "Function Acceptance."

WHAT IS WORTH?

The original value analysis formula presented difficulties in measurement. The denominator (cost) is directly measurable in units of U.S. dollars (or yen, marks, pounds, or any other currency). But the numerator—worth—is not precisely measurable. The dictionary definition—Webster's Ninth New Collegiate Dictionary states, in part, that "worth is the value of something measured by its qualities or by the esteem in which it is held . . ."—is of little help.

Neither "the qualities" nor "the esteem" is directly measurable, yet our equation requires that they be measured. This forced the early value analysts to compromise the process. The system that emerged, called basic function value, can be paraphrased as follows: worth . . . the lowest cost to obtain the basic function (Miles 1961).

This oversimplification misled a generation of value analysts. It implies that the measurement of the worth of an item is a simple matter of defining a single basic function for the item and then summing the cost of the parts of the product that perform that basic function. This basic function cost is then compared with the total cost of the item. If they are very different, the item is regarded as having low value.

This exercise in the definition of what was called the "worth" of the basic function is a useful motivator in a value analysis training seminar, but it has limited validity in the real world of users and customers.

Thomas J. Snodgrass identified this problem in the 1950s, when he was manager of engineering of General Electric's Hotpoint Division. His solution was a system for indirectly but accurately measuring true worth of an item by directly and accurately measuring the reaction of actual user/customers.

Modern value analysis draws heavily on the Snodgrass system. Prior to any value analysis team effort, data is gathered that reflects both the needs and the wants of the typical user or customer. It is then possible to assess the degree to which an item fulfills those needs and wants.

Chapter 5 details techniques for measuring users' wants and needs and converting those measurements into indices of function worth.

VALUE ANALYSIS OBJECTIVES

In the original General Electric workshop format, the projects to be value-analyzed were identified by management or a value analysis department. Neither group was truly qualified to identify projects that represented low value.

In modern value analysis, the projects to be value-analyzed are identified by the value analysis workshop team. This assures that any such project represents a value problem that needs to be solved.

It is important to distinguish between projects and products. The product to be value-analyzed is, indeed, identified by management, based upon business-related criteria. The projects to be attacked are identified by the team, based upon an entirely different set of criteria:

Group	Criteria
Management (in identifying the product to be attacked)	Is the product the heart of the line, in that improvements will spin off to many other products?
	Is the product important to the future of the organization?
Value analysis team (in identifying the projects to be value analyzed)	For which functions do the function costs fail to match the user's needs and wants?
	When focusing on these targeted functions, what projects are identified for the creativity and synthesis phases?

VALUE ANALYSIS OR COST REDUCTION?

In what way do these value efforts differ from the good old-fashioned cost-reduction systems that are an essential element of any successful business?

The answer is, in two very significant ways—the user focus and the function orientation—as shown below:

	Cost reduction	Value analysis
The formula	Value = cost	Value $= \dfrac{\text{worth}}{\text{cost}}$
The questions	What is it? How can we make it for less?	What is it? What does it do? What must it do? How can its functions be performed better or for less?

Value analysis aims to provide the user-required functions at lowest cost. Cost-reduction systems aim at reducing the cost of the present parts.

Early-Stage Value Analysis

Much value analysis is performed on existing products. A significant proportion of value analysis, however, is performed at the design concept stage. The returns from such early-stage effort are far greater, since the analysis precedes expenditures for planning and tooling, and the results of the value analysis are implemented in all units produced. In addition, the power of value analysis is made available to guide the rational development of the product.

Value Control

The original objective of value analysis was to reduce the cost of existing products. In the late 1950s, Lawrence Miles's value services group at General Electric conceived a new system called value control. Its objective was to plan for and control product cost to meet the required product profitability of the enterprise. GE elected not to publicize value control to the outside world as it had with value analysis. They saw it as a unique and fundamental restructuring of the new product development process, giving them a significant competitive advantage. Only recently have the details of this advanced system been made public, by Robert L. Bartlett, the GE engineer whom Miles assigned to develop and implement value control.

Value control is solidly based on value analysis principles, with major modifications to deal with the problems of new product development and decision making.

One of the key concepts of value control is the market standard. It is correctly assumed that the prices paid in the marketplace for a class of product are a reliable measure of the worth of its functions and functional specifications to the user/customer. The market standard is translated into a market standard cost, which is the maximum the product may cost if the producer is to meet the profit objective.

As with modern value analysis, the value control process starts with the creation of a costed diagram of all the functions performed by the product. This shifts the viewpoint of the design engineer toward the purpose or mission of the product and away from its hardware.

The value control process then shifts the focus to competing products. The costs of each significant competitor are allocated to the same function structure. This data, coupled with a systematic search for design alternatives, yields the value standard for each function, defined as the lowest cost to provide a function that meets the essential specification. The power of the search for value standards as a means of generating significant product value improvements cannot be overstated.

The value control system was specifically developed to improve the appallingly high rate of "infant mortality" among new products. It also provides a reliable system for modernizing and improving the profit potential of existing products. The system was tested on a wide variety of GE products, with impressive results.

When applied to a new product's development, value control usually bases its analysis upon existing products of the same class. This assures that the plans arrived at for the product are consistent with the market standard concept and will meet the customer's basic needs and wants.

Value control is a rigorous and powerful technique. It requires the dedication of substantial resources over an 8-to-12-month period. The effort required is justified wherever there are vigorous competitors and a desire to leapfrog over competition.

Comparative Value Analysis

Modern value analysis has adapted one of the elements of value control: that two or more dissimilar items that perform the same task can be directly compared at the level of their functions.

A controlling assumption of the comparative value analysis process is that to succeed, a product must be no more than 15 percent new. This was proven valid even on that archetype of a unique product, the original "printshop-in-a-desk," the Xerox 914 copier.

Although modern value analysis of a single product invariably results in dramatic improvements, the full power of value analysis is unleashed when the process is applied simultaneously to two or more different products that perform the same user task.

When a product has been converted to a FAST diagram, it is possible to compare directly the functions of each element of the product on a one-to-one basis with any other product or process that accomplishes the same purpose.

This capability, through function analysis, to compare directly every part of, for instance, a Mazda rotary engine using lobes and tip-seals with a Pontiac using cylindrical pistons and rings, promotes a previously unattainable insight (Figs. 2–2 and 2–3).

This insight has benefited value analysis systems in such diverse areas as offset printers, electric motors, and clothes washers.

The same techniques permit comparative value analysis of two or more products from dissimilar producing areas—for example, piston compressors versus screw compressors, the overhead costs at two different locations, and the production process at a main plant compared with a branch plant.

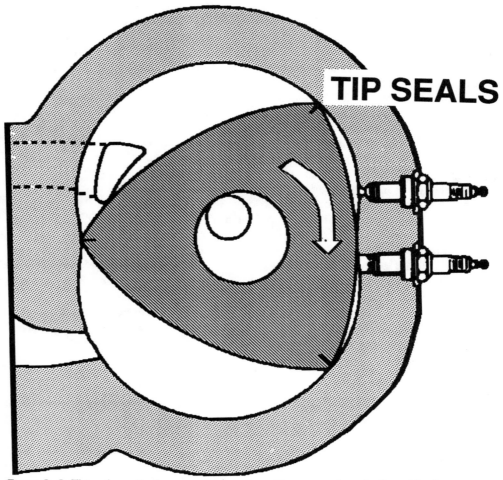

Figure 2–2. This schematic shows a rotary engine with apex seals on its three-lobed rotor.

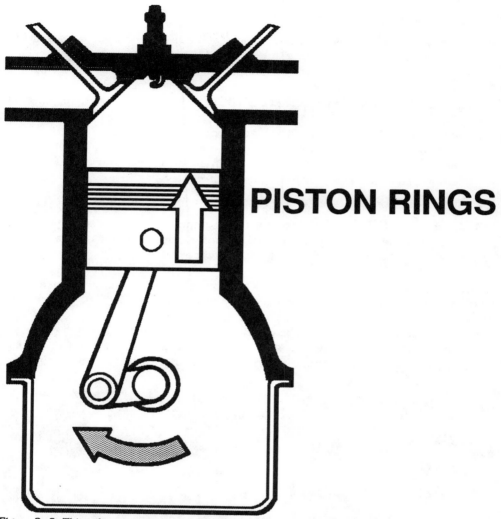

PISTON RINGS

Figure 2–3. This schematic shows a reciprocating engine using cylindrical piston rings.

In some companies, comparative value analysis is used to benchmark how well an existing product is doing against the competition, or to establish functional targets for future products.

The power released through comparative value analysis stems from this fundamental precept of Miles-ian value analysis.

If, in the pursuit of better value, functions have not been identified and those functions have not been evaluated by comparison, then the process has not been value analysis, but merely cost analysis (Miles 1961, 15).

The Value Analysis of Burden and SG&A

The majority of value analysis is product-oriented; that is, it is applied to the material and labor content of a product that will be delivered to a user or customer.

The most dramatic value analysis studies, however, attack those items that are commonly refered to as burden, overhead, or SG&A (selling, general, and administrative expense). The single term overhead is used in the discussion below to represent all of these accounting artifices.

There are three primary reasons for the power of such an analysis:

1. These costs are allocations, that is, are grouped together and allocated to an account. This process serves to disguise their identity.

2. They are seldom effectively analyzed except to compare differences in cost ratios.

3. The overhead value analysis team usually comprises the top operating manager and his or her staff (Fig. 2–4). The function viewpoint permits these decision makers a new and uncommonly focused viewpoint.

The users and customers of overhead are primarily those within the organization who regularly interact with and benefit from the activities funded by the overhead accounts.

Typical accounts that constitute the overhead cost include indirect payroll, warranty, overtime payroll, utilities, indirect material, industrial engineering, small tools, manufacturing engineering, shop supplies, purchasing, receiving, material control, shipping,

Figure 2–4. A management team works on a FAST diagram during value analysis of overhead.

quality assurance, equipment rental, tool maintenance, equipment repair, and supervision.

Savings on overhead costs through value analysis can be impressive. In one example, a pair of high-level teams at a major automotive company, dividing the accounts in accordance with their areas of specialization—one from manufacturing, the other from the front office handling primarily non-manufacturing decisions—produced after two months $2.56 million in implemented reductions, or 14.2 percent of total overhead costs.

Value Analysis in Public Utilities

Telephone, gas, or electric companies and other public utilities are unique economic entities. They must return a profit to their stockholders, but those profits can be constrained by the actions of a state public utilities commission. One would think that such an environment would not nurture value analysis. It would appear to be a classic case of separating the source of funding from the source of authority, removing all motivation to improve efficiency or increase profit.

Nevertheless, a number of public utilities have established effective value analysis systems. The Indiana Gas Company in Indianapolis initiated its system, at the suggestion of a particularly creative member of the Public Service Commission, with three successful value analysis efforts on the billing system, strategic gas operations, and underground gas storage. It has since established a continuing program with regular and impressive results, chiefly due to the informed involvement of a pair of executives, the senior vice president of operations and engineering and the vice president of engineering.

Chapter 3

THE VALUE ANALYSIS JOB PLAN

THE ORGANIZED APPROACH

Derivation and Definition

Since its creation by the Greek scientist Anaxagoras, the scientific method has guided philosophers, scientists, and engineers in solving problems as vast as the theory of universal gravitation and as mundane as how best to prevent the spoilage of canned fruit.

Value analysis is both a philosophy and a problem-solving system.

It is, however, primarily a procedure: "a series of steps followed in a regular definite order," as defined in Webster's Ninth New Collegiate Dictionary.

The classic scientific method procedure has nine steps:

- Decide on objective

- Analyze

- Gather data

- Organize data

- Induce

- Plan

- Precheck

- Activate plan

- Evaluate

Miles chose this as his model for the value analysis job plan. His initial plan had six phases:

- Information

- Speculative

- Analytical

- Program planning

- Program execution

- Status summary and conclusion

In the 40 years since the job plan was conceived, there have been hundreds of variations, most of them developed in an attempt to personalize the program. Though they vary in length from 4 to 14 steps, they all contain the following essential core: an information phase, creativity or speculation phase, and analysis or evaluation phase.

Though these three phases cover the central power and uniqueness of the value analysis system, we have chosen an eight-step version to emphasize the central elements and the equally essential supporting elements.

The modern value analysis job plan is shown in Figure 3–1, moving from the bottom step upward.

The original value analysis job plan was commonly applied in a 40-hour, one-week training environment. The modern value analysis job plan is not for training. Its objective is problem solving. To accomplish this objective typically requires 55 hours in session over an eight-to-ten-week period. A very complex study may require the addition of 1 or 2 days to the schedule, stretching the in-session hours to 63, or even 71. The typical schedule is shown in Table 3–1.

The Ideal Creative Problem-Solving System

The power of the value analysis job plan derives from its precise match to the three-step ideal creative problem-solving system: (1) Load the mind, (2) divert the mind, (3) create a receptive environment. The result is a new combination of known facts—a creative idea.

Creativity, most modern behavioral psychologists agree, is the development of a new combination of known facts. In *The Act of Creation,* Arthur Koestler describes the apparent process by which this combination comes to pass.

We define it as an apparent process because no one knows how the mind processes data. Even Koestler's evocative description of the process is an allegory, though based upon empirical observations.

Koestler defines the creation of a new engineering design or scientific discovery as "the permanent fusion of matrices of thought previously regarded as mutually incompatible" (Koestler 1964). If the fusion of two distinct planes of thought creates a viable concept, we have a new idea—perhaps a scientific discovery or a unique engineering design (Fig. 3–2).

The elegance of Koestler's concept is inspiring. By defining this process, he is describ-

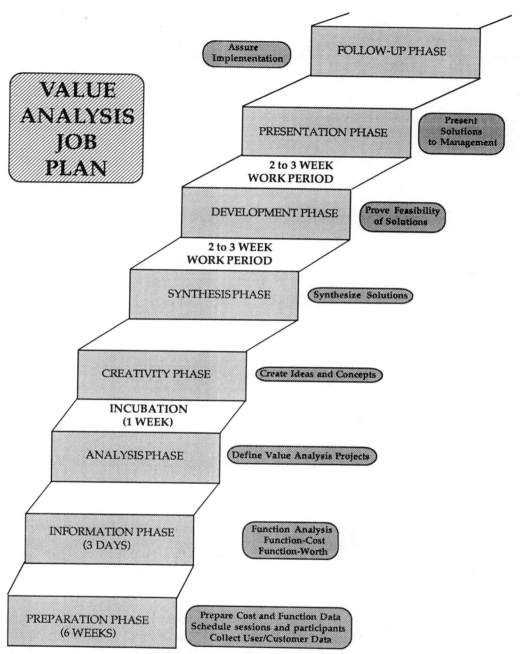

VALUE ANALYSIS JOB PLAN

FOLLOW-UP PHASE — Assure Implementation

PRESENTATION PHASE — Present Solutions to Management

2 to 3 WEEK WORK PERIOD

DEVELOPMENT PHASE — Prove Feasibility of Solutions

2 to 3 WEEK WORK PERIOD

SYNTHESIS PHASE — Synthesize Solutions

CREATIVITY PHASE — Create Ideas and Concepts

INCUBATION (1 WEEK)

ANALYSIS PHASE — Define Value Analysis Projects

INFORMATION PHASE (3 DAYS) — Function Analysis / Function-Cost / Function-Worth

PREPARATION PHASE (6 WEEKS) — Prepare Cost and Function Data / Schedule sessions and participants / Collect User/Customer Data

Figure 3–1. The eight steps of the modern value analysis job plan are depicted in the form of a rising stairway.

TABLE 3–1. TYPICAL SCHEDULE FOR A VALUE ANALYSIS WORKSHOP STUDY

Activity	Period	Phase
Prepare for program	Start six weeks prior to first team meeting	Preparation
Focus panel or questionnaire	Before team meetings	(User input)
Function definition	First workday	Information
Function costing	Second workday	Information
Function worth	Third workday (A.M.)	Information
Identify value analysis targets	Third workday (P.M.)	Analysis
(Several-day break for unconscious incubation of ideas)		
Create words/ideas	Fourth workday (A.M.)	Creative
Elevate to concepts	Fourth workday (P.M.) and fifth workday (vendors/experts in P.M.)	Synthesis
(Three-week period—members verify championed concepts)		
Refine concepts	Sixth workday	Development
(Three-week period—members complete championed proposals)		
Sell value analysis proposals	Seventh workday	Final presentation
Implement proposals	Until implemented or rejected	Follow-up

ing the functioning of an entity that no one has ever seen—the subconscious or unconscious mind.

Indeed, the unconscious mind does not physically exist at all. The concept is really no more than a convenient way of referring to an observed set of phenomena. Unconscious acts are simply those acts over which a person has no conscious control.

Koestler states, and in his book goes on to prove dramatically, that no problem was ever solved *creatively* in the conscious mind (Koestler 1969).

The only way we solve problems in our conscious mind is through recalling stored facts. That is, we can consciously recall previous solutions to problems. We cannot, however, consciously interrelate two facts and define a unique combination of the two. This interrelationship or intersection can take place only in the unconscious mind, where we have, by definition, absolutely no control. It is for this reason that creative solutions often seem to pop into a person's mind unexpectedly.

Inventors recognize loading the unconscious mind with pertinent facts as an essential step in the creative problem-solving process (Fig. 3–3). According to Nobel Prize winner Ivan Pavlov, "Perfect as is the wing of the bird, it never could raise the bird up without resting on air. Facts are the air of the scientist. Without them you can never fly" (Osborn 1953, 141). Alex Osborn, creator of the brainstorming process, refers to this mind-loading as "prospecting" (Osborn 1953, 143). He describes his favorite techniques in terms of "meandering through stores" or flipping through the Sears Roebuck catalog.

When Koestler's thesis was published in 1968, it was a burst of clarity to those studying or working with value analysis and the creative process. Koestler's theories describe the fundamental process of creative problem solving. The terminology, the procedures, and the sequence are precisely duplicated in the value analysis job plan.

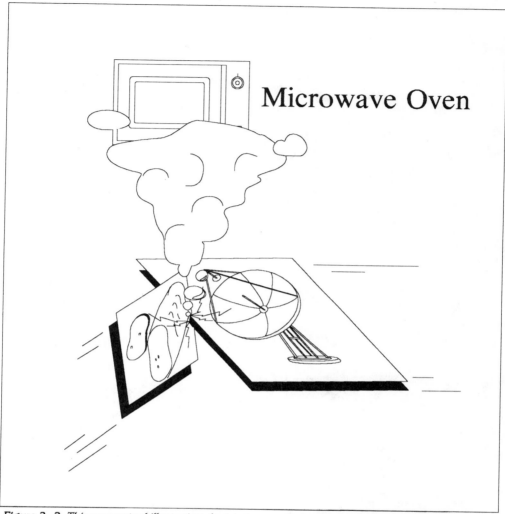

Figure 3–2. This conceptual illustration depicts two separate planes of thought—cooking and the power of satellite-dish transmission—intersecting to create a new frame of reference: the microwave oven.

Value analysis, as it empirically evolved at General Electric, was simply an effective system of problem solving. Koestler's work reveals that modern value analysis is far more: It emulates the ideal creative problem-solving system.

Adaptation to Value Analysis

Since unique solutions can be created only in the unconscious mind, there are really only two things we can do consciously to facilitate generating them: 1) we can fill our unconscious mind with all the elemental facts; and then 2) shift our focus away from the problem in order to permit our unconscious mind to feed back the creative solutions.

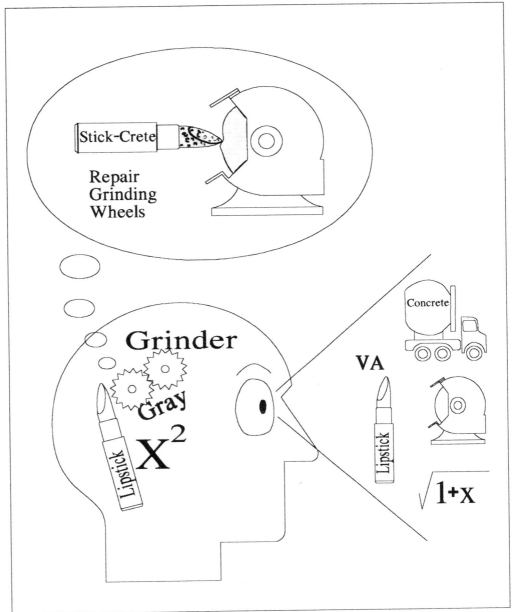

Figure 3–3. Like all other true creators, team members load their minds with data and then combine them subconsciously, with creative results.

Those on the leading edge of value analysis have been instinctively led by these concepts in their effort to optimize the process over the past 30 years. The result is modern value analysis.

Guided by the value analysis job plan, we first fill the unconscious mind by forcing a high-tension review of only the pertinent facts: those involving function, cost, and user

attitude. Further enhancing the process is our treatment of the facts about function in a form that distracts attention from the fixations of present hardware, by defining these functions in an elemental form called the *functive*.

We then permit several days for unconscious incubation.

Finally, we establish a judgment-free environment called a creative session, during which the unconscious feeds back the solutions. Figure 3–4 summarizes these three steps.

Harry Rosenfeld, the highly effective leader of the Xerox Corporation's value analysis system, has suggested that this three-step process emulates the excursion concept of the powerful creative problem-solving system of Synectics®, described by *Fortune* magazine as "inventing by the madness method" (Alexander 1965).

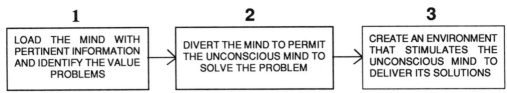

Figure 3–4. The optimum creative problem-solving system comprises three separate steps.

Plan for Implementation

The first step in the value analysis system is to prepare a detailed plan for the implementation of any conceivable proposal that the team may develop, using the implementation planning worksheet form shown in chapter 6 (see Fig. 6–5). This is in keeping with the famous Leonardo da Vinci exhortation: "think of the end before the beginning."

The objective of this step is to anticipate the inevitable roadblocks that will be raised in the path of implementation, and to take early action to disarm objections. This effort also develops within team members an early sensitivity to the difficulties that face them in their pursuit of change.

This *anticipative* technique recognizes a number of industrial realities:

Most proposals for change are rejected.

The reasons for rejection most often surprise (and frequently irritate) the proposer.

These reasons for rejection, with minor exceptions, could have been anticipated. Many could have been resolved in the early stages of the problem-solving process.

They are not anticipated simply because people generally believe that "if it's a good solution, it will be implemented. For many reasons, this is simply not so.

The problems involved in implementing change have been generally recognized by those who advise us on how to find solutions. Their systems, however, invariably list a step called "plan for implementation" at the *end* of the problem-solving process.

The error in this approach is that if we wait until we have solved the problem before we develop a plan for implementation, it is too late to take many of the actions called for in the plan!

The "Theme Thread" Company and Product

It is possible to describe value analysis as a series of concepts or viewpoints. We have chosen instead to create a "Theme Thread" company and product line (Fig. 3–5), which will be used throughout this book as a way to illustrate the process with a sense of reality and immediacy.

"SURE, BUT WILL IT WORK ON MY PRODUCT?"

It would be presumptuous to expect that the reader will freely and readily accept the precepts of modern value analysis as a replacement for his or her present methods for solving problems. It is common for a person being introduced to the value analysis

COMPANY NAME	COMPANY SIZE	
General Corporation	**ANNUAL SALES:**	$30 Million
Memphis, TN	**EMPLOYEES:**	2,000

VOLUMES

PRODUCT LINES	ANNUAL QUANTITY	CUSTOMER/USER BASE
Heat Pumps	12,000	$16 Million (10% direct to Contractors for refurbishing. 90% direct to Contractors for new construction.
MIG Welders	3,500	$3 Million (15% direct to major manufacturers, shipbuilders, construction, etc. 85% to professional welders through 1800 distributors.)
Dehumidifiers	80,000	$11 Million (90% private labeled to Sears and Penney's. 10% to smaller labels.)

PRODUCT SELECTED FOR VALUE ANALYSIS

Model 4700, 4-Ton Rooftop Heat Pump, Sideflow input, downflow output, gas supplemented.

COST (MLO):	$1060
PRICE:	$1378
QUANTITY:	Presently 5,800 per year

RATIONALE FOR PRODUCT SELECTION

The probability is high that changes in this model will be applicable to other Heat Pumps in the line. If 25% of the changes apply to the remainder of the line, the yield from cost reduction alone would exceed the savings on the target model by a factor of over 2.

The 4-Ton model is the heart of the General Corporation's Heat Pump line. Its labor and material content are typical of 53% of total Division product.

The entire General Corporation Heat Pump line is highly accepted in the market, being regarded generally as the standard against which competitors are compared. It is maintaining its 20% market share even in the face of a 5% to 28% price disadvantage.

TARGETS

Cost reduction of 28% after elimination of all non-tolerated faults.

RATIONALE FOR TARGETS

SIZE OF MARKET:

For 4 Ton: 150,000/year, this year
135,000/year, last year
110,000/year previous year

MARKET SHARE AND PRICES:

General Corporation has 4% of the market. Stable for 5 years.

Major US competitor is 5% lower in price and has 18% market share which has dropped 2% in last 2 years.

Korean competitor is 28% lower in price. Has come from nothing to 15% market share in 2 years.

Eight other competitors share 55% of market. All are lower in price than General Corporation by 5% to 20%.

Figure 3–5. This is a detailed description of the "Theme Thread" company and its products, which will be used for reference throughout this book.

techniques for the first time to listen appreciatively for a while, then react with a statement such as "I can see that it worked on the example you are describing, but my product (or process) is different. It won't work for me!"

The "Theme Thread" in this book is very specific. It thus presents many opportunities for the reader to take exception. For example:

"The General Corporation is a $30 million company. We do $300 million. It won't work here."

"They produce institutional products for contractors. We sell a consumer product. It won't work here."

"They produce a hardware product. Our product is computer software. It won't work here."

The reader must translate each of the steps of the modern value analysis system into his or her own frame of reference. Remember that this process has been proven effective in an unlimited variety of environments; it succeeds wherever the problem involves both cost and function. Throughout your study of modern value analysis, it is necessary only to remain open-minded.

Chapter 4

THE PREPARATION PHASE

LEARNING BY DOING

Value analysis cannot be taught through classroom lecture. It cannot be completely learned simply by reading about it in a book.

Value analysis can be considered learned only when the participant has changed his or her problem-solving behavior.

This requires total involvement by the participant in defining and solving problems and implementing solutions.

The value analysis team workshop carefully nurtures a structure and an environment that permit—in fact, force—this total involvement of each of the participants.

Structure

The team rigorously applies the value analysis job plan, with off-site meetings for approximately 55 hours over a period of six to eight weeks. In addition, each team member will apply approximately 60 hours to the verification of championed proposals during two three-week break periods.

Environment

The prime requirement for an effective value analysis team workshop is reality. To that end, several constraints apply:

1. The objective is defined in terms of a product, that is, a physical item that the organization produces (or plans to produce) or a service or procedure that it provides (or plans to provide) to a defined user/customer base.

2. The product (or planned product) is both current and real.

3. The product to be value-analyzed is identified by top operating management.

4. Top operating management is the involved leader of the value analysis system.

5. Participants are the experts on the product who would have been assigned to develop or improve it even if there were no value analysis.

6. Data on function, cost, and user/customer acceptance are both valid and current.

7. Where it has been decreed that the value analysis system will be applied only to part of a product or service, the team will first analyze the total product and then focus its efforts on the designated part. This assures that user needs and wants will be the primary control throughout the process.

The Preparation Process

Before a team begins its value analysis work, much needs to be done, from deciding on the focus of the study and makeup of the team to gathering all the data required (Fig. 4–1). The preparation process includes the following elements:

- Product selection
- Selection of participants
- Required technical data
- Required cost data

Figure 4–1. A value analysis team defining the functions of an architectural roof window.

- Required equipment and logistics
- Required market data
- Required user data

The value analysis team must be freed from unmotivating drudgery. Preparations must be so complete that the session can concentrate on value analysis, rather than on such ancillary tasks as data collection and cost calculations.

The process of preparation described in this chapter and the next should be followed rigorously. Schedule a full six weeks to complete the preparation.

PRODUCT SELECTION

The projects on which the value analysis team will focus its creative problem solving are identified by the team on the third day of the workshop.

Initial identification of the specific product that the team will analyze is made by top operating management, based on such factors as design cycle status and importance to the future of the organization. The value analysis study will be most effective if the product chosen is the heart of the line, that is, the product with significant future promise and with maximum potential for spin-off to the remainder of the product line. The product chosen must be a specific model to ensure that the study maintains a focus on specific costs and specific features rather than on diffuse generalities.

For example, Figure 4–2 shows the entire line of food mixers that the Hobart Corporation produces. The product chosen for value analysis in this case was the ten-quart

Figure 4–2. *The entire product line of Hobart food mixers, ranging from 5 to 200 quarts.*

mixer, which the company judged to be its key long-term product, the model with the greatest potential for spin-off to larger and smaller models.

SELECTION OF PARTICIPANTS

This is not a training program; it is a problem-solving system. Do not merely fill the slots with people who happen to be available. Select team members whose knowledge and job assignment best equip each to contribute to valid product improvements. Some examples of the composition of an ideal five-member team are given in Table 4–1.

A team with the proper balance of commitment, competence, and stimulation is the most critical factor for a successful value analysis workshop. Creating such a team can be accomplished by limiting team size, choosing participants from different functional areas, and selecting those with the ability and authority to evaluate and implement proposals for change.

Ideal Size

Experience has proven that five-member teams are the most effective. With four or fewer members, there is usually insufficient interaction or cross-fertilization of ideas. With six or more, the sessions take on some of the characteristics of committee meetings—members fail to interact or function merely as advocates of their own viewpoints.

Avoid the use of a pickup team, that is, a team having members who are assigned to participate simply to obtain experience in the practice of value analysis.

Figure 4–3 illustrates the benefits of cross-functional teams through the principle of the interaction window. This optimal interaction results from the overlap of a variety of areas of knowledge and competence in the makeup of the value analysis team. The team must be multidisciplinary. It comprises carefully selected decision makers whose collective experience is appropriate to the product or process being analyzed, and whose background and interests are conducive to constructive interaction.

Great care must be taken in selecting a team with the proper balance of differing backgrounds and personalities, but with a common frame of reference. We suggest that the team membership be completed in an interactive session with the key functional managers and the value analysis coordinator. Strongly resist any last-minute changes or substitutions.

The cross-functional nature of the team is one of the keys to effective value analysis.

TABLE 4–1. AN IDEAL VALUE ANALYSIS WORKSHOP TEAM

Component	Function	Examples
Design	The "engineer's engineer"	Project engineer, chief draftsperson, designer; ideally, the engineer responsible for the product under study.
Operations	The doer	Industrial engineer, manufacturing engineer, process/methods engineer, factory supervisor
Cost	Cost data	Cost estimator, industrial engineer, accountant
Outreach	User-input	Marketing, sales, field service, purchasing
Catalyst	Stimulator	A constructive skeptic, possibly from engineering, product management, or marketing

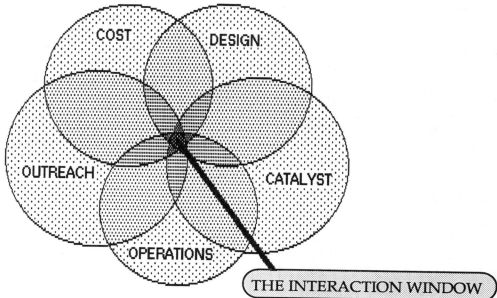

Figure 4–3. This Venn diagram depicts the interaction window within which the value analysis team will perform—the ideal overlap of five organizational capabilities that are required on a value analysis team.

It is not uncommon, however, for a team to include more than one member from a specific area of responsibility. It is possible, for instance, that the ideal catalyst comes from the same functional area as another team member. This causes no serious problem—except in one case: The team must never include more than one knowledgeable design engineer. The effectiveness of the five-person value analysis team depends strongly on the absence of any centers of power; each team member is a minority of one. The presence of two or more engineers would represent a block of expertise that would tend to intimidate the others, making them less likely to challenge any statement on which the two engineers agreed.

Level of Expertise

The basic rule for team member selection is that they all be key employees. Most should be decision makers or experts in their functional areas. Ideally, each team member should be a person whose present assignment includes responsibility for the product under study.

The Key Man

Early value analysis teams tended to exclude from membership any person with key technical knowledge and authority on the product under study. It was felt such a person would resist change, preventing the team from reaching beyond what was already proven and implemented. This was supported by much early experience with project engineers on teams, whose favorite phrase seemed to be, "We tried that. It didn't work."

It is true that the project engineer will be protective of his or her design. The other team members are not engineers, and the conventional reasoning is to wonder, "How could they possibly contribute" in an arena that is so clearly beyond their capability and experience?" Nevertheless, modern value analysis requires that team membership include either the project engineer or another key person with intimate technical knowledge and authority involving the product under study. This person—by convention, regardless of gender—is referred to as the key man, the most technically knowledgeable person available; ideally, the person with the authority to sign any engineering change order.

This key man is required on the value analysis team simply because in the end, little or no effective change can be accomplished without his or her involved support.

Team Leadership

The key man is not necessarily the functioning leader of the team. During the first value analysis study session, team members democratically elect a captain, preferably someone who is relationship-oriented rather than simply technically knowledgeable. When selecting team members, management should take into consideration the inclusion of a person who meets this criterion.

Equality on the Design Team

Despite the terms *key man* and *team leader*, it is a crucial responsibility of the value analysis coordinator to blend together the disparate members of the team so that each will accept his or her role as an equally qualified participant on a design team.

This requires that the project engineer (or other key man chosen as the technically knowledgeable member of the team) accept that each team member, whether an industrial engineer or comptroller or purchasing agent, is capable of fundamental contributions to the design of the product under study.

It also requires that those same members accept their coequal role; the team will fail if members regard themselves as merely advocates or protectors of their own professional viewpoints.

When all team members accept this, the invariable result is a series of unique solutions, often triggered by team members who didn't know their idea wouldn't work, and nurtured by a key man who is willing to listen.

It is occasionally found that five otherwise ideal candidates have less than compatible personalities. In such a case, some organizations find that a two- or three-day team building course is useful in welding the team into an effective value analysis instrument.

Number of Teams

It is occasionally desirable, in order to spread the workload on a complex product, to assign two, three, or even—where circumstances are just right—four teams to the effort. The possible reasons for such an arrangement include a very complex product (one with more than 1,000 line items of cost, including material entries and labor operations) and a desire to expose more than one team to the value analysis process to increase the pool of people experienced in it.

When this happens, it is appropriate to segment the product, assigning to each team an equal proportion of the line items of cost. The teams are structured to match the segmentation of the product with members chosen for their areas of specialization. The basis for the segmentation varies with both the product and the capabilities of the team members. In one case, an earth mover was segmented into main frame and cab; boom, carriage, and powertrain; hydraulics and power optimization; and miscellaneous. A two-team value analysis of a bank security system was segmented into mechanical and electrical areas.

In a multiteam value analysis, all teams work on the whole product. The segmentation applies only during the function-cost effort on the second day. This focuses the analytic talents of all team members on areas with which each is most familiar.

There is a tendency for each such team's final proposals to relate to its assigned segment. It should be strongly emphasized, however, that all participants are simultaneously studying the entire product.

To assure that each team in a multiteam study retains a cohesion of purpose and approach, regular melding sessions are held throughout the value analysis study.

Simultaneous Studies

The effectiveness of a value analysis effort is greatly increased when two or more products are studied simultaneously. Although it is impractical to assign two products to any one team, up to four teams may study up to four products in the same room.

The limit of four teams is based on the monitoring capabilities of the value analysis coordinator. The workshop staff is commonly increased by making use of previous value analysis workshop study participants when more than one team is involved. Even with a staff of one experienced value analyst per team, however, no more than four teams should be permitted. When this limit has been exceeded in the past, the personal interaction between the coordinator and the participants has been lost, resulting in a failure to accomplish a crucial objective: Every participant must become convinced that value analysis is the most effective system available. This failure often returns some nonbelieving participants back into the mainstream of the organization, where they will inevitably poison the value analysis system and compromise its effectiveness.

The interplay generated in a multiteam study benefits all of the teams. This is in addition to the obvious improvement in the value analysis coordinator's efficiency.

Team Member Commitment

The members of a value analysis team must commit themselves to 100 percent participation throughout the study. Part-time participants can seriously compromise study results. Whenever it is determined that any member is either failing to attend sessions or not applying the required intersession effort, the value analysis coordinator should dissolve the team and reschedule the study.

REQUIRED TECHNICAL DATA

Once the team members have been selected, each must be supplied with a detailed agenda and a list of participants including job title, location, and phone number.

Each team must be supplied with a technical package consisting of the following items:

Indented, costed bill of materials and labor

Engineering bill of material, specifying materials, sizes, quantities, and so on

Complete set of labor routing sheets

Complete set of drawings, including purchased parts. (If more than one team is working on the same project, each is provided with the drawings of only its assigned segment of the product/process.)

Data on the product or process being studied and on all significant competitive products or processes

Two sets of actual hardware or, for a process, complete flow charts and exhibits of the process. (For hardware, one set should be assembled, one disassembled.)

Information on purchased parts and materials, such as prices, total annual usage, vendor sources, tooling, and any problems.

Each team must also be provided with market data, including product history (customer acceptance, the competitive situation, and problems), market factors, and user attitudes (either a complete set of questionnaires or the full-focus panel report); a list of experts and vendors, with a description of products or specialties; and a cost estimating guide. Details on this information are given below.

Price and Cost

Throughout this book, the word *cost* appears prominently. Cost and price are totally separate matters.

Cost is defined as the sum of the labor, material, and burden dollars that the producer invests in the product.

Price is defined as the number of dollars that a buyer will pay for the product.

It is important in value analysis to prevent any mind-set that regards price simply as cost + markup.

REQUIRED COST DATA

Cost is one of two factors that determine value, as shown in the following equation:

$$\text{Value} = \frac{\text{worth}}{\text{cost}}$$

Most cost systems are structured to run the business and are not commonly in a form that permits a team to analyze each product or process element.

The teams must therefore be supplied with true and detailed cost data on their project.

Disadvantages of Raw Cost Data

It is possible for a team to function using only a bill-of-material structure, a listing of the costs of purchased materials and a set of labor routings. The disadvantages of using data in this raw form, however, should not be ignored.

First, the team is distracted from its assigned task of solving problems by being required to perform unmotivating drudgery in the form of multiplying labor factors, cross-referencing piles of documents, and so on. Members' efficiency and accuracy are commonly lower than those whose usual responsibilities include this sort of data manipulation. Further, excessive time is wasted, often adding extra days and cost to the value analysis effort.

Best Cost Data Structure

The ideal data format, shown in Table 4–2, is called the indented, costed bill of materials and labor (ICBOML). Any similar format is acceptable, if it conforms generally with four requirements:

1. Each element of material or each operation of labor is entered on a separate line.
2. All costs are in terms of dollars (not hours or minutes).
3. Variable labor overhead (fringe only) is included in the labor cost.
4. Costs are shown both for each item and extended (multiplied by quantity per product).

If such a specialized cost package cannot be prepared for the teams, it is possible, albeit more difficult, for the teams to function with a bill of materials, with the labor costed at labor and fringe, and a set of labor routers, also with the labor costed at labor and fringe.

Cost-Estimating Guide

A brief guide is provided that will permit team members to prepare spot cost estimates during the synthesis and development phases. Information about a heat pump produced by our "Theme Thread" company provides an example of a cost-estimating guide (Fig. 4–4). Required data for such guides typically include the following:

Annual sales volume for the specific "model" of the process or product under study

Annual sales volume for any related "models" for which there is a possibility that changes to the "model" under study may also apply

A format that will guide the uninitiated in estimating cost by simply entering such elements as material cost and standard hours and then performing some simple mathematics

Standard material costs per pound and per cubic inch for all raw materials that are, or could be, used on the product being studied

TABLE 4–2. INDENTED, COSTED BILL OF MATERIALS AND LABOR

Motor, starter, P/N 30A417   8,000/year

| Part or operation | | Qty. | Mtl. $ | Labor + OH | | Total | (notes) |
Number	Name			Hrs.	$		
143A3432-2	Housing/field assembly	1	$2.4653		10.6045	13.0698	
op 010	Assemble complete	—		.0882	2.0461	2.0461	
32F477	Housing, motor	1	1.2240		1.8548	3.0788	
42M321	Tbg, crs, DOM, 5" od	8.0"	1.2240			1.2240	($.153/")
op 010	Machine pilots	—		.0249	.6718	.6718	
op 020	Drill (5), tap (4)	—		.0497	1.1830	1.1830	
131F423	Coil, field	2	.4538		5.4082	5.8620	
66M114–16	Wire, magnet, #16	0.6#	.3738			.3738	($.623/#)
94M320–2	Tape, insulating, 1/2"	32"	.0800			.0800	($.0025/")
op 010	Wind on mandrel	—		.1026	2.3393	2.3393	
op 020	Form, bake	—		.0405	.9234	.9234	
op 030	Tape coil, strip (2)	—		.0941	2.1455	2.1455	
33F101–12	Pole piece	2	.6251		1.2954	1.9205	
22M140	Armco ingot iron 3/8 lg	1.465#	.6251			.6251	($.4267/#)
op 010	Forge to print	—		.0260	.6188	.6188	
op 020	Drill, tap, c'sink (2)	—		.0254	.6766	.6766	
29191–3816	Scr, FH, Stl, 3/8–16	4	.1624			.1624	($.0406/ea)

PRODUCT UNDER STUDY:
   MODEL 4700 Heat Pump

QUANTITY FORECAST:

MODEL	CURRENT YEAR	YEAR +1	YEAR +2	YEAR +3
4700	5800	6500	7500	9000
3820, 3830	2500	2500	-0-	-0-
5000 Series	3800	4500	5500	5500

MATERIAL COSTS:

1020 STEEL	$ .32/#	$1.28/in^3
380 CRES	$1.70/#	$6.07/in^3
ABS PLASTIC	$1.80/#	$ .07/in^3

LABOR RATES:

DIRECT LABOR
   Grade 1 $14.20/hour
   Grade 2 $12.10/hour
   Grade 3 $10.10/hour

INDIRECT LABOR
   Grade 1 $11.20/hour
   Grade 2 $10.30/hour

ENGINEERING
   Design 1 $22.80/hour
   Design 2 $19.50/hour
   Test $17.60/hour

DRAFTING
   Designer $16.20/hour
   Draftsperson $13.40/hour

MFG ENGINEER $21.50/hour
TOOL DESIGN $14.20/hour
TOOLMAKER $12.50/hour

LABOR OVERHEAD, ALL LABOR (FRINGE ONLY)   **38%**

TOOLING AMORTIZATION   **1 YEAR**
CAPITAL AMORTIZATION   **3 YEARS**

MANUFACTURING COST = STD HOURS X LABOR RATE/HOUR + STD MTL
SCRAP FACTOR = STD MTL X 3%

Figure 4–4. This example of a cost-estimating guide is for the "Theme Thread" company's heat pump.

Direct and indirect labor rates for all categories of internal labor, including implementation labor such as engineering design, drafting, test, manufacturing engineering, toolroom, and so on

Labor overhead rate, consisting of only the fringe benefit adder

Standard amortization rates for tooling and capital

A statement of the organization's make-or-buy policy

Scrap factors

REQUIRED EQUIPMENT AND LOGISTICS

Physical preparations for the value analysis workshop should be sufficiently thorough that the team is able to maintain full concentration on the value analysis system. Appendix B includes a list of the equipment and logistics items that must be in place (see Fig. B–2).

REQUIRED MARKET DATA

Each team's data package typically includes a summary of three factors relating to the market into which the particular product or process fits.

1. Annual sales: Actual figures for similar products or previous designs; predicted usage for the product or process under study.

2. Market share: Results of any market research predicting potential usage by year and by market area. Present and predicted shares of this potential achieved by this organization and by each significant competitor.

3. Competition: Data on all significant present and potential competitors.

UPSTREAM VALUE ANALYSIS

The power of modern value analysis in identifying and removing unnecessary cost from existing products is legend. Its power with new products at the early concept stage is demonstrated regularly and dramatically, but is far less well known.

The reason the value analysis literature seldom mentions upstream value analysis is that such a project lacks a "before" for the usual comparison. The early-stage product is not yet even formally designed, and so we have no so-called base case in the form of drawings and costs. Thus, it can be difficult to prove that the product has benefited from value analysis.

While some value analysis has been attempted on concept-stage products with no drawings or cost data, such systems invariably degenerate into simply creative exercises, for the following reasons:

1. It is impossible to develop a valid user-oriented FAST diagram in the absence of a specific and detailed design.

2. Without detailed drawings, the value analysis team is isolated from the intimate function relationships that are the source of the revelations so typical of a successful value analysis.

3. Without detailed costs, the previously stated value equation cannot be quantified.

4. Without a "before" to compare with the "after," there is no measure of the effectiveness of the value analysis effort. It is thus essential that a base case—a set of drawings or sketches and a detailed cost estimate—be prepared before the start of a value analysis study.

The Base Case

As previously pointed out, value cannot be determined without knowing the detailed present cost of the product. In upstream value analysis, we not only have no costs, but also have no drawings or hardware. We are therefore faced with the challenge of creating a set of detailed drawings and a detailed costed bill of materials for a product that may be no more than a concept in the mind of a project engineer or product manager. At first glance, this task appears to be impossible.

Note, for example, a common development cycle in U.S. industry (expressed in number of months):

0	Product concept proposed
3	Start of research and development
6	Concept approved
12	Laboratory model successful; start design
30	Design complete; release to manufacturing
36	Costs first defined

It certainly appears to be a difficult challenge to define the detailed design and the detailed costs at the time of concept approval, 24 to 30 months before such data is typically available.

In fact, however, it is a rather simple process, one that should be part of *any* design process, even one that does not include value analysis. It is merely necessary to accept two fundamental assumptions and then to allocate a limited amount of manpower and schedule time.

Assumption 1: Even at the earliest concept approval stage, within the collective minds (and the desk drawers) of the project engineer and the key designers and draftspeople are essentially all of the details that will later be formally placed on the drawings. Admittedly, these represent only the status of the design concept at this early stage. The thousands of decisions the project engineer will field during the development process have not yet been made. Indeed, the testing and calculation upon which these decisions will be based have not even been performed. It is, however, this already available data that top management judged to be sufficiently valid to permit concept approval.

Assumption 2: No more than 15 percent of the elements of any new product design are actually new to the organization. The remaining 85 percent of the elements of the design are either identical to portions of existing products or are modifications of existing elements.

This results from the clear pressures of competitive business logic. A classic example of the validity of this assumption is the Xerox 914, the original Haloid Corporation development of that revolutionary concept: a desk-sized print shop. The elements of its detailed design were less than 15 percent new. The remainder, even those elements concerned with Xerographic transfer, were chiefly modifications of designs being produced by others. (It should be noted, however, that the marketing of this dramatic creation was, of necessity, nearly 100 percent new.)

Defining the Base Case

This process requires a base case team of two to four people for one to four weeks, depending on the complexity of the product. The base case should be defined before the concept is submitted for top management approval.

The project engineer is charged with the responsibility of creating a complete set of assembly and detail drawings of the design concept as it exists at the time, in the collective minds of the design group. These drawings may be formal sketches, marked-up blueprints, modified CAD drawings, or simply quadrille pad sketches. Some will undoubtedly be taken without change from the current drawing files. Each receives a unique sketch number. A parts list ties them into a structured drawing package. Those items that simply cannot be detailed to this degree, even in this unconstrained format, are described on a separate sheet in as much detail as practicable. Designer and draftsperson support is applied as needed.

The base case team also includes one or more industrial, methods, and/or process engineers, working closely and in parallel with engineering. As a detail sketch is completed, it is "planned" and costed down to the labor operation level. Reference data (machine type, labor classification, assumptions) are added to the face of the sketch. Materials are costed by purchasing. An indented, costed bill of materials and labor (ICBOML) is created.

A relatively simple product requires 1 to 2 man-weeks, while a very complex system may require four people and 16 man-weeks over a one-month period. This is clearly an investment in money, in manpower, and in schedule. There is commonly an initial reluctance to make this investment, for two reasons. First, the project engineer knows the ultimate design will be different, with many months of changes in the initial concept, based on an unlimited series of trials, tests, and decisions. Experience advises this engineer that the design should not be published until this design process is complete and the product is proven to perform its required function. In addition, management is generally reluctant to permit the expense of preparing such a drawing set and cost set, because it has not seemed to be necessary in the past.

Experience has proven, however, that when management is apprised of the need for such data for effective value analysis and when the project engineer realizes the great value of a well-defined base case to guide design decisions, both enter into the process enthusiastically.

It is occasionally possible to minimize the preparation effort by defining a base case using data from a similar product, produced either by the organization or by a competitor.

Chapter 5

THE MEASUREMENT OF USER/CUSTOMER ACCEPTANCE

USER INPUT

The prime focus of modern value analysis is on satisfying the needs and wants of the user of a product. The teams must ensure that these needs and wants are fulfilled before they attack any unnecessary costs.

Needs are relatively easy to handle. They are usually controlled by specifications and are the result of a series of objective agreements between the user/customer and the provider. Wants are harder to handle. They are subjective. It is difficult to set limits on such characteristics as convenience, satisfaction, and appearance.

As a result, many value analysis systems ignore customers' wants as too difficult to measure and even more difficult to specify.

A method that identifies value analysis targets, however, serves to focus the problem solving of modern value analysis on the fulfillment of both user needs and wants.

A function that is identified as a value analysis target is simply one in which the user need or want does not match the cost of that function. The objective is to focus creative effort and later synthesis effort only on functions where such a match does not exist; that is, areas where improvements will increase user acceptance and/or reduce cost.

The only effective way to determine what the user/customer needs or wants is to ask.

Two acceptable methods are available for defining user needs or wants for a product or process: 1) the open-ended questionnaire and 2) the user focus panel.

Both are effective. The determination of which to use in a specific set of circumstances is based on three factors:

- The type of product or process
- The characteristics of the various users or customers
- Available resources

THE QUESTIONNAIRE METHOD

This method uses the expertise of professional marketing research organizations to perform personal interviews (either by telephone or face-to-face) of a relatively small, highly focused sample of users/customers (Figs. 5–1 and 5–2). Its advantages include a large volume of responses and in-depth analysis by a properly trained and motivated interviewer who can read the nuances of the responses. This often leads to further verbal probing and deeper insight. The major disadvantage of the questionnaire method is a higher cost than the user focus panel.

Sample Size

A typical valid sample is 30 to 50 respondents per product. This relatively limited sample is sufficient because of the highly focused nature of the respondent selection criteria and the open-ended nature of the inquiries.

Qualification of Respondents

Approximately one-third of the respondents should be employees of the producing organization who have a user's viewpoint, such as decision makers from marketing, field service, or quality assurance. At least one-third must be outside users. The ideal user has made the buying decision within a year or so and has used the product—or a directly competing one—for several months. The remaining respondents are outsiders who have experience using a competing product and are potential users or customers.

Figure 5–1. A professionally trained interviewer conducts research by telephone.

Figure 5–2. Carrying a clipboard, an interviewer conducts face-to-face research with a user of a rider mower.

The source of outside respondent prospects is commonly the marketing department. Warranty cards or sales records or simply the other records and knowledge of salespeople are useful resources.

Form of Interviews

While the ideal format is the face-to-face personal interview, a properly conducted telephone interview can yield totally acceptable data.

Interview Documents

A questionnaire similar to that shown in Figure 5–3 is prepared jointly by the value analysis coordinator and the marketing department. It serves equally well for the face-to-face or telephone interview.

The structure of the document is open-ended, in that the major emphasis is on the probe, that is, the unconstrained comments that the respondent offers following certain types of questions: "Are there any more reasons?" "Why else do you like it?" "Anything else?" "Any further concerns?" "Give me some examples" "What else is important?"

Questions are also "salted" into the questionnaire to gather data unrelated to value analysis. It is common for marketing or operating management to request that this opportunity be used to collect user data on other specific concerns. The team does not use this data in its analysis phase effort to define value analysis targets, though these answers often contribute to the team members' store of user-related information.

The structure should avoid placing the respondent in an artificial environment. Avoid the format of the so-called mall encounter, in which the respondent is asked, for instance, "If you were buying a Mercedes, would you buy a blue one or a red one?" Such questions will always elicit a response, and there is no way to determine whether the response is valid.

DATE: *August 3, 1993*

PRODUCT: *Model 8700 8-Ton Gas/Electric Rooftop Heat Pump*

MANUFACTURER: *General Corporation*

NAME OF RESPONDENT:
Alex Founder

LOCATION: *Baltimore, MD*

QUALIFY RESPONDENT:

☐ IN-HOUSE _____
(Title and area of responsibility)

OUTSIDER

☒ USER OF PRODUCT BEING ANALYZED *7*
(months of use)

☒ USER OF COMPETITIVE PRODUCT *Carrier 45L8085*
(manufacturer of product)

☐ DEALER/DISTRIBUTOR _____
(Heat Pump product Lines)

SPECIFIC JOB AND RELATIONSHIP TO THE HEAT PUMP *Building-owner and operator of a Central Laundry processing Building.*

(1) Why do you own a General Corporation Heat Pump? *I like the efficiency.* **+3** *Good reputation for long life.* **+5**

(probe) *General is a good name.* **+5**

(2) Overall, how would you rate the effectiveness of this Heat Pump?
☐ Extremely effective   ☐ Fairly effective
☒ Very effective **+6**   ☐ Not particularly effective
    -2 ☐ Not effective at all
(probe) *Very Efficient. Only problem was a slight leak in piping. Fixed under warranty. (Probe) Problem was a badly soldered joint.* **-2**

(3) Is there anything about the operation of the Heat Pump that annoys you?
YES ☒ NO ☐ (probe) *Noisy.* **-3** *Vibrates the roof over the office area. Not that bad though. (Probe) Maintenance man says it's Hell to disassemble for repair,* **-6** *but so far it's so dependable that it doesn't* **-15** *matter. He also says you need a whole tool kit to repair a General. No two fasteners are alike.* **-3**

(4) Do you have any complaints about the size or configuration of the Heat Pump?
YES ☒ NO ☐ (probe) *Well, we did have to put in extra roof beams to hold it.* **-3** *The physical size doesn't matter though. It's on the roof!*

Figure 5–3. *This sample of a user-attitude questionnaire asks about the "Theme Thread" heat pump, and it includes encircled responses and ratings of their importance and seriousness.*

(5)   How about power consumption? _____ It's the *most efficient unit on the market.* **+5**
That's important to me.   **+5**

(probe)   *The Energy Miser feature is great!*

---

(6)   In general, how much room for improvement do you think there is in the way the Heat Pump performs?

☐ A whole lot   ☐ A fair amount   ☒ Only a little   ☐ None

---

(7)   Has anything worn out or deteriorated on your Heat Pump? *Only the leak* **-2** *And it's rusting on the top, but that doesn't really matter. It's on the roof, remember?*
-2

(probe)   *From my experience, General makes the most dependable unit around* **+5**

---

(8)   Are there any features of the Heat Pump that you hardly ever use? _____
☒ YES   ☐ NO   (probe)   *The side output feature. It seems to me that there's a lot of cost in that, and I don't need it. My unit is installed with the output on the bottom.* **-2**

(probe)

---

(9)   If a good friend asked you whether to buy this Heat Pump, would you say . . .
☒ Buy it   ☐ Don't buy it   ☐ Don't know

---

(10)   What, in your opinion, is the most important consideration in selecting an 8-Ton Heat Pump? _____ *Dependability, then Efficiency. (Probe) Also ease of repair.*

(probe)   *I want the unit to just sit on the roof and run. (Probe) I think it should last 20 years with only minor maintenance.*

---

(11)   Is the Heat Pump easy to maintain? ☐ YES ☒ NO   (probe) *When it does break down it's hard to take apart. When the electronics breaks down, you have to replace the whole control box.* **-2**   **-2**

(probe)   *My maintenance man likes Carrier. It's not quite as dependable, but when it does break down, it's easier to work on.*

---

(12)   If you had to make your buying decision over again, would you buy this Heat Pump?
☒ YES   ☐ NO   ☐ Don't know

---

(13)   Reflecting on your entire experience with this Heat Pump, would you say that it works the way you expected it to?

☐ Much better   ☐ Worse
☒ As expected
☐ Better   ☐ Much Worse

Thomas F. Cook
(interviewer)

Figure 5–3. Continued.

The Interviewers

This interviewing method is usually a new experience to people in marketing research organizations, both in-house and outside, who have not previously worked with this form of user-oriented value analysis. They must be actively trained by a person familiar with the modern value analysis system.

Analysis of the Heat Pump Questionnaire

The 19 usable responses in the sample completed questionnaire shown in Figure 5–3 have been circled, classified, and rated in a joint session between the value analysis coordinator and marketing.

The number of usable responses will range from 5 or 10 (from a reticent respondent) to a more typical 25 to 40. If 30 respondents yield an average of 30 usable responses each, the data base will contain 900 responses, sufficient for the value analysis team to make valid judgments later on product worth.

The rules for annotating a questionnaire at the joint session are simple but rigorous. Each distinct response element that truly represents the respondent's reaction to the product under study is circled. If a response element is repeated, it is circled each time it appears.

After the response elements are circled, each is classified by attaching a + to each element that represents a like and a − to each element that represents a dislike. Then each + or − is rated on a scale of 1 to 10. A rating of 5 indicates that the element is fairly important. A rating of 1 indicates an element of minimal importance, while 10 indicates an element the respondent regards as most important.

The rating process is essentially a subjective one. Its objectivity can be improved, however, if the raters consider each response within a larger framework, that of the aggregate body of questionnaire response data. The raters read through all of the responses to gather an overall impression of the respondent's frame of reference. They then rate each response within this framework.

In making rating decisions, the rating questions (numbers 2, 6, 9, 12, and 13 in Figure 5–3) are useful to establish a collective image of the point-of-view of the respondent.

If there is insufficient information to identify the importance of an element, it is arbitrarily rated a 5.

Note that a number of the respondent's comments in Figure 5–3 are not circled. These either were determined to be not significant to the purposes of the study, are merely expansions of specific thoughts already recorded, or are comments that do not relate to the specific product under study, though they may be of interest to marketing. It is not uncommon for significant marketing information to emerge from a carefully crafted, uninhibited questionnaire effort.

THE FOCUS PANEL METHOD

When properly planned and conducted, the focus panel method will generate sufficiently valid data for the value analysis team to establish value analysis targets. Its prime disadvantage is that it requires substantial in-house effort to identify and schedule appropriate participants. The effectiveness of the panel is almost totally dependent on the proper selection of panel members and the session leader. It has two major advantages: (1) Its cost is sharply lower than the questionnaire method, and (2) it actively involves key marketing, sales, and operating managers, resulting in greatly improved believability of both the data and the ultimate results of the value analysis study.

It is especially important that there be total agreement between the value analysis organization and marketing management. The two viewpoints that must be effectively melded are (1) the value analysis objective, which is to provide data to the teams to permit them to define value analysis targets in the analysis phase, and (2) the organization's marketing objectives, which are often constrained by a preexistent marketing plan.

Qualification of Focus Panel Members

The ideal focus panel has about 15 members. The ideal mix is six to eight outside "user/customers" and a roughly equal number of key management decision makers from the organization that designs, produces, and markets the product.

Outside Panel Members

A comment by Tom Peters and Nancy Austin in *A Passion for Excellence* exemplifies how to choose the outside members of the focus panel: "While 'average' customers may reflect yesterday, **lead users**—i.e., forward-looking customers—are usually years ahead of the rest . . ." (Peters and Austin 1985, 120).

The outside panel members should be carefully selected by the marketing department, with the close involvement of the value analysis coordinator. They should be lead users, people whose actions will significantly affect the future of the product. They should reflect the key market segments of the future. They should also include some people who affect the purchase decision only through their influence on others. At least one of the panel members should be, in the opinion of marketing, a prime buying influence.

In-House Panel Members

The seven members of the producing organization whose opinions are required include people with the following titles or their equivalent:

- vice presidents of marketing and engineering
- product manager of the product or process under study
- national and regional sales managers
- manager of field service
- project engineer for the product under study (who typically is also a member of the value analysis study team; see chapter 4)

These titles may vary when the product under study is a process or a service rather than a hardware product.

It is not appropriate, in most cases, to include operations or purchasing on the panel unless the person performing that function also fits one of the above categories. The objective of the focus panel is to gather unconstrained data on the product's worth in the market. Inclusion of operating personnel from within the organization tends to blur the data. The place for these operating people is on the value analysis team.

The participation of executive or top operating management on the panel is very appropriate, except where such an authoritarian presence would compromise the objectivity of the other participants.

Preparation for the Focus Panel Meeting

Each participant should be informed that the meeting will require approximately six hours, including lunch. Part-time attendance is unacceptable.

The objective of the focus panel is to sense the wants of the members on a real-time basis, without any prestructuring. No advance thought or preparation is necessary. Indeed, participants are discouraged from preparing lists of likes or dislikes in advance of the session.

Format of the Focus Panel

Participants are seated at a conference table with actual user/customers on one side. On the opposite side are the members of the organization (marketing, sales, engineering).

It begins, for example, with a welcome and introduction by the vice president of marketing, from 9:00 to 9:10 A.M. The value analysis coordinator then defines the value analysis system and establishes the rules. At about 9:30 A.M., a session leader assumes responsibility for the remainder of the session. This person should have market research skills and should be an outsider, not a member of the organization.

The Role Panel Members Play

Each outside member of the focus panel is instructed to respond throughout the session from his or her own frame of reference.

The in-house panel members, however, should play a distinct role. They will respond throughout the session from the frame of reference of the prime buying influence of the product under study. A brief session is held at the outset to define this prime buying influence. It is pointed out that every product or process has many users and customers. For example, those for an automobile include the owner, the owner's spouse, any other driver, any passenger, mechanics, and perhaps a fleet owner; for a heat pump, they include the dealer/distributor, contractor, building owner, architect, and repair technician. The other key point made is that one of these user/buyer/customer categories represents the prime influence on the buying decisions for a significant share of the future sales of the product under study.

Focus panel participants first list all candidates for prime buying influence, then select one of the candidates as the most effective role for the in-house participants to play. The focus panel then spends the rest of the morning identifying and ranking user/

customer likes. After lunch, the focus panel identifies and ranks dislikes. At about 3:00 P.M., the participants rate each of the key competing products on each of the previously identified likes. This is followed by a free and unconstrained listing of "heart's desires" of all of the actual or potential users of the product.

The procedure for each of these three inquiries is simple and elemental. For the likes session, the leader simply repeatedly asks the question, "What do you like about (the product under evaluation)?" The objective is to extract the participants' true viewpoint about the product. Their unconstrained reactions are the window through which these viewpoints appear. The leader takes no part in the identification of the likes, simply guiding the discussion of each expressed viewpoint until it appears that all of the panel members understand the viewpoint as it is expressed. The leader then prepares for the voting by asking the panel these two questions: (1) Is this viewpoint indeed a feature of the product under study? and (2) do all members of the panel understand the viewpoint as it is read back by the team recorder?

The Voting

In-house participants are instructed to respond throughout the session as they feel the prime buying influence would respond. Outside user-participants are instructed to respond from their own frame of reference.

Each panel member then simultaneously displays his or her rating of the importance of the like. The vote is recorded. This continues until the panel runs out of positive comments. The leader repeatedly asks, "Are there any more things you like about (the product under study)?" When there is no response from the panel for several seconds, the leader declares that all significant likes have been defined.

The session then moves into its second phase: the identification of dislikes. The process is identical. The much-repeated question is, "What do you dislike about (the product under study)?"

Dislikes are rated somewhat differently from likes. The panelists are asked to consider two factors: the consequences of the failure or complaint, and the probability that the failure or complaint will occur.

The seriousness of the dislike is then voted.

All voting takes place simultaneously with the display of Olympics-style voting cards (Fig. 5–4) or through the use of an electronic display voting system such as the Q-system (Fig. 5–5). The range of allowable votes is 1 to 10. The number 7 is often not used, because it has proven historically to be a "safe haven" number; disallowing its use forces the participants to commit themselves.

Each vote is recorded, electronically or on a spread sheet, which also carries the final wording of the like or dislike as voted.

As an example of this procedure, Figure 5–4 shows 14 user/customers and managers evaluating a wiring system called Trenchduct®, a product of Butler Manufacturing's Walker Division in Parkersburg, West Virginia. The voting cards that the panel is displaying represent each participant's judgment of the seriousness of a fault, or dislike, that they have just identified and discussed.

The outside user members of the panel represent a cross section of the prime buying influences. Each was chosen as typical of the architects and electrical contractors who specify the wiring systems for new construction.

Figure 5–4. A focus panel votes by displaying numbered cards, as judges do in the Olympics.

The Focus Panel Report

A report is prepared for distribution to the value analysis workshop team during the analysis phase. Each viewpoint is shown exactly as it was worded for the vote. The mode of the vote—the most frequent rating—is calculated and recorded. (Note that the mode is used, not the mean or the median, since the mode represents where the vote grouped; that is, where there was group consensus. The mean and the median, for the purpose of assessing behavior, are nothing more than mathematical artifices.)

Figure 5–6 is an excerpt from data collected at a focus panel on the "Theme Thread" product, the 8-ton heat pump.

Benefits of This Approach

Note that likes and dislikes are identified by the panel, not by management or marketing or the value analysis coordinator. This dramatically improves the validity of the responses and believability of the results.

Another benefit of this focus panel approach is the heavy involvement of management, particularly the marketing, sales, and engineering managers. As a result of their early involvement in the value analysis system and their active participation in the creation of the focus panel data, the road is smoothly paved for the value team. When the team later presents results to management and marketing, the results tend to be highly believable.

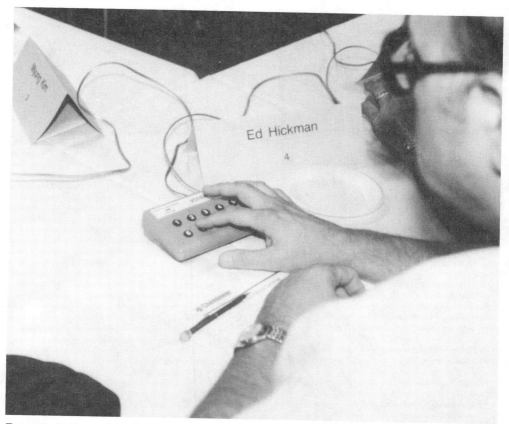

Figure 5–5. A member of a focus panel sits in front of a Q-System pushbutton pad. The electronic device is manufactured by Reactive Systems, Inc., of Englewood, New Jersey.

Marketing often views this deep involvement in the value system as a unique and welcome opportunity to affect the product fundamentally.

It is common for a number of the likes or dislikes to relate to concerns other than the product under study, such as company policy or practices of sales, distribution, or service. These are recorded, discussed, and voted for the consideration of both marketing and top management.

Competitive Evaluation

The third phase of the focus panel is an attempt to determine the collective opinions of the panel members about the degree to which each of the key competitors satisfies the users' wants with respect to each of the likes defined in the first phase. This data is not essential to the definition of value analysis targets. It is, however, of great value to the team in its design effort, since it represents the individual and collective viewpoints of both the managers and the actual users about the relationship of the present product to its market.

The method is simply to read aloud each like and then, if it can be rated measurably,

DESCRIPTION OF FEATURES (OR "LIKES")	IMPORTANCE OF FEATURE												NOTES	MODE & RANGE	FUNCTION
	MANUFACTURER							USERS							
	1	2	3	4	5	6	7	8	9	10	11	12			
43 Less than 1% compressor failure in 5 years.	8	8	6	8	5	8	8	8	10	3	4	9		8R7	
44 Low Silhouette (less than 3 feet).	5	9	8	9	8	8	6	4	6	8	6	8	(4)	8R5	
45 Supply and return sides coincide with Carrier.	6	10	10	8	8	9	10	10	9	8	10	10	(4)	10R4	
46 Easily field convertible from side output to down.	10	10	10	10	10	8	9	9	10	10	10	5	(4)	10R2	
47 Single point power, factory installed.	9	10	9	9	8	10	10	10	10	9	9	9		9R2	
48 Ability to complete sheet metal work to curb before unit arrives.	9	9	10	10	8	8	9	10	10	8	10	10		10R2	
49 Ability to rig without spreader bars.	4	8	8	10	9	10	6	10	10	10	3	10		10R7	
50 Parts interchangeable between models.	8	9	8	9	8	10	10	9	10	10	8	8		8R2	
51 Thermal expansion valves versus capillary tubes.	8	8	4	4	3	4	4	5	9	6	9	6		4R6	
52 Dealer/Contractor diagnostic capability	6	9	-	6	5	10	9	4	8	9	10	4		9R6	
53 Crankcase heater.	9	10	8	8	6	5	6	8	10	5	9	9		8R5	
54 Energy Management Controls available as option.	3	4	4	-	8	2	5	4	6	9	3	4		4R7	
55 5 min. delay on compressor as standard	6	10	6	8	6	9	6	6	8	8	10	10	(2)	6R4	
56 Concentric transition within curb.	2	3	-	-	3	1	-	0	6	1	-	1	(4)	1R5	
57 Power vent on gas-heated unit.	4	6	4	8	5	3	4	9	8	3	9	6	(3)	8R5	

Note (1) Multimodal - Rating shown is the MEAN
Note (2) Multimodal - Rating shown is the MODE nearest the MEAN
Note (3) Data spread is excessive and invalid; Rating shown is the MEAN
Note (4) Not allocable; Does not relate to the specific model under study

Figure 5–6. This sample data sheet from a focus panel report describes the voting on several likes about the "Theme Thread" heat pump.

use the voting cards or buttons to rate the product and its competitors on how well each performs that specific feature.

It is not necessary in this competitive evaluation process to rate more than two competitors; the "high side" and the "maverick." Try to choose them on the following basis: Choose a competitor that is regarded as the standard of the industry. If the product under study is that standard, then select the competitor that is most admired; that is, the competitor whose product you would be proud to produce if your organization did not produce the study subject. Choose as a second competitor, if possible, one that is regarded as a maverick, a nonconformist, making products so different that your engineers may tend to dismiss them as second-rate—"We would never produce that kind of product here!" Marketing has reported, however, that this competitor is increasing market share and beginning to affect sales.

Applying the Focus Panel Data

Whether the user/customer's attitude is defined through a questionnaire or a focus panel, the data is submitted to the value analysis workshop team in the analysis phase. The process of allocation of this data to the FAST diagram is detailed in chapter 6.

THE INFORMATION PHASE

THE DESIGN TEAM MIND-SET

By definition, the value analysis workshop team is a diverse group. Each of the five members comes from a different area of specialization. All are decision makers and prime experts on the product under study. This would seem to be a formula for anarchy —and indeed, without a concerted effort at mind-setting, such a team does tend to degenerate into five different problem-solving groups, each with a membership of one and each promoting its own viewpoint.

Mind-setting does not take place as a discreet activity. It is not a separate phase of the value analysis job plan. Rather, it is a framework established during the six weeks preceding the team's first meeting and continually reinforced throughout the team study. It has four key elements:

1. Rigorous selection of team membership.

2. Delivery of the workshop manual, with reading assignments, two weeks before the first session.

3. A two-hour briefing for all team members during the week preceding the first session.

4. Continuous reinforcement of the study's design team environment.

The single objective of mind-setting is to convince all participants that it is to their advantage to function as subordinate team members rather than as proponents for their areas of specialization—as receptive searchers for alternatives rather than as protectors of the status quo.

Recognize that with such a carefully constituted team, the protective instincts of the team members are nearly always positive impulses; they are commonly protecting something they feel might be degraded by "less qualified" team members. It is an honest instinct. The challenge is to sublimate or redirect these impulses to a common, team-directed purpose.

Team Balancing

As the first element in the mind-setting process, a panel composed of the operating management and led by the top operating manager selects the team members. Although such top-level attention may seem presumptuous, it is appropriate to the size of the investment of resources in the value analysis study and to the importance of the product under study. It is also appropriate to the size of the probable return on that investment and to the potential positive effect of a value analysis success on the future of the organization. Finally, such top-level involvement ensures attention to the critical objective of team balancing.

The team selection session starts with a 15-minute briefing by the value analysis coordinator, describing the five capabilities required in an ideal team (see Table 4–1). A matrix is then drawn on the board (Fig. 6–1).

Slots are filled for each team in an interactive session that the value analysis coordinator leads. As each slot is filled, team composition is tested for technical competence, constructive discontent, compatibility, and stability.

In most sessions, the panel will conclude that more data is required in order to make a final decision on one or more of the selections. Appropriate panel members accept the responsibility to obtain information and report back to the top operating manager for final selections.

Reading

Delivery of the workshop manual to each team member two weeks before the first team session is the second element in the mind-setting process.

Team members are instructed to read the preparation, orientation, and information phase chapters before their scheduled briefing.

[BOARD OR FLIP-CHART]	**TEAM NUMBER**			
	1	**2**	**3**	**4**
DESIGN				
OPERATIONS				
COST				
OUTREACH				
CATALYST				

Figure 6–1. The team selection matrix is drawn on a board or flip chart to help the top operating manager and panel select team members.

Briefing

In the focal point of the mind-setting process, all team members meet with the value analysis coordinator for a two-hour briefing (Fig. 6–2). The message of the briefing is that the targets established for this effort are far-reaching; the team members cannot restrict themselves to considering only high-probability design concepts, but must reach beyond what they have previously considered to be normal bounds. They must, however, be reassured that the value analysis system will reestablish these normal bounds before their results are completed and presented to management.

Each team member must leave the briefing believing that he or she is a member of a design team, not simply a proponent or protector of a particular area of specialization. Reinforcement of this team spirit is the fourth key element of mind-setting.

The team should also conclude by the end of the briefing that an appropriate team captain is not necessarily technically oriented, but preferably relationship- and team-oriented. The value analysis coordinator should announce that each team will select its own captain on the first meeting day.

THE SUCCESSFUL VALUE ANALYSIS WORKSHOP

In the past, value analysis has been performed by individuals. It has also been performed in the classroom teaching environment of a value analysis workshop training seminar. It has been performed by engineers, purchasing agents, accountants. Since the early 1950s, hundreds of thousands of people have performed value analysis—and the majority of it has failed to fulfill its promise completely.

Figure 6–2. The value analysis coordinator, standing at the flip chart, briefs team members.

The promise of value analysis is, quite simply, that the miracles performed in the often-quoted early GE workshops can be repeated by anyone, simply by following the job plan. Most value analysis fails to repeat those miracles for the basic reason that most value analysis does not emulate those early GE workshops. What drove those workshops to their dramatic accomplishments? What features characterized the early workshops that Lawrence D. Miles and Roy Fountain led in Schenectady, New York?

Key Characteristics

Those early workshops had three distinguishing features. First, all team members were carefully selected decision makers. Second, all team members were highly motivated, universally proud to have been selected for this pioneering effort. They were motivated by their sense of being special, much like those famous Hawthorne Works groups at Western Electric in the early 1900s. Third, teams were most often responsible for seeing their recommendations through to implementation. Members identified with their proposals, in effect becoming champions for those changes.

In addition to these three distinguishing characteristics of the workshop, the team members applied what were known as the "five value analysis questions," the six-phase "value analysis job plan," and the "13 value analysis techniques."

Modern value analysis has established a structure built upon those early miracles but revitalized by 30 years of experience and experimentation. For value analysis to succeed as in those early workshops, you must follow this structure with great rigor.

The Keynote

It is essential that the top executive or top operating officer present a keynote address to kick off the value analysis workshop.

There are two key objectives of this critical five-minute presentation. The first is to demonstrate graphically not only the top level support for the effort, but also, and much more important, informed involvement with the value analysis process. The second objective is to define a specific and measurable goal for each team. This goal must be rationalized in terms of the specific objectives of the organization with respect to its market.

The guide shown in Appendix B (see Fig. B-7) lists key points to be made in the keynote presentation.

The Implementation Plan

To avoid later surprises, the team prepares, as the first step in the information phase, a list of all possible areas of change. Consider, too, what possible sources of opposition might develop to each area of change listed on this implementation planning worksheet. Determine how this opposition might be defused and record in an adjacent column several actions that should be taken to ensure implementation of each change. As the study progresses, the team regularly refers to this list, taking action to minimize opposition as well as updating the lists of areas of change and of actions required to defuse opposition.

If the team has anticipated effectively, it will find, upon reaching the synthesis phase, that the road to implementation is considerably less cluttered. Ideally, for any change proposal that reaches the final presentation phase, there will be no surprises.

The Team Leader

Each team is instructed to select one of its members as a captain to lead it through the value analysis process. The election is purely democratic, but members are urged to avoid selecting their most technologically knowledgeable member. They are encouraged instead to elect a person whose behavior is primarily relationship-oriented. People whose focus is on promoting personal interaction will tend to unleash the practical creativity of the team better, blending the implied authority of the engineer with the less constrained search-oriented viewpoint of other team members. The selection of a captain should take place late on the first day of the workshop to permit team members sufficient time to observe each other.

FUNCTION ANALYSIS

The functional approach to problem solving is the cornerstone of value analysis. It is the single unique element of the remarkable system of problem solving that Lawrence D. Miles developed in the 1940s.

Function analysis translates the structure (parts and labor processes) of any product or process, procedure or service, into a structure of words. The objective is to shift the viewpoint of the problem solver from the concrete to the unconstrained.

It converts the labor operations and the materials into verb and noun statements. The team might, for instance, focus on optimizing the two-word combination "extend life," rather than the far narrower goal of "reducing the cost of the lubrication system."

The value analysis team members, who are all experts on the problem to be solved, know the study product intimately. They know the labor operations of the service or product. They know the materials from which the product is made, and the process by which it is made. In short, they are the most knowledgeable people available on the product under study.

This is both an essential advantage and a curse. The expert "knows" all of the possible solutions to the problem. Indeed, he or she has often tried many of them. It is difficult for such a person to back away from the realities of the product as he or she knows it and accept that fundamental improvements are indeed possible.

Function analysis solves this problem. A function diagram is a total semantic equivalent of the product being analyzed. Every labor operation, every material element is represented by one of the verb-noun combinations. The expert, having helped to create this new language of the function diagram, is confident that analyzing only the functions is the complete equivalent of analyzing the product or process itself.

FUNCTION ANALYSIS SYSTEM TECHNIQUE (FAST)

In 1963, a Univac logician named Charles W. Bytheway conceived a new method of function definition and analysis. He named this method function analysis system technique —FAST, for short—and saw it primarily as a stimulus to creativity. Bytheway used seven thought provoking questions, each of which must be answered using only two words, a verb and a noun. These two-word combinations, the functions performed by the product, are termed *functives*. They are the heart of all value analysis. Bytheway arranged these two-word combinations on a diagram in a how-why relationship, forming what he called a FAST diagram.

A FAST diagram is essentially a structuring of functives into logically interrelated groups. This grouping of functives is the key to the entire process.

The past 20 years have seen significant improvements in Bytheway's dramatic concept. As a result, two major forms of FAST are in current use: technically oriented FAST, used when the product being analyzed has no identifiable customer or user, and user-oriented FAST, the form used in modern value analysis.

The FAST diagram is a precise semantic equivalent of the product under study. In Figure 6–3, the FAST diagram contains 42 independent functions. Every piece of material and every labor operation related to the earth excavator under study contributes directly to one or more of those functions. One of the excavator's functions was lengthen life. Bytheway's "how" question resulted in seven additional functions, entered to the right of lengthen life, including resist damage, reduce friction, eliminate contaminants, and control temperature. Although the value analysis team concentrates only on the functions, it is examining the total excavator. Its viewpoint, however, is not that of material and labor, but rather the far less constraining viewpoint of function. The team concentrates its problem solving on what the product does, rather than what it is.

Figure 6–3. This photograph shows a completed FAST diagram of an earth excavator under study.

Creation of Functives

All definitions of functions must be expressed in only two words, a verb and a noun. Where possible, the verb in the functive should be demonstrable on a nonverbal level, and the noun should be a measurable property. Figure 6–4 lists numerous suggested verbs and nouns.

Note that the term *functive* is not a universal replacement for the word *function* in value analysis activities. It is used specifically to refer to or emphasize the form in which functions are expressed. *Function* is normally the accepted term—except where it is essential to emphasize the nature of the functive as the two-word essence of the function.

When creating functives, avoid passive or indirect verbs such as *be, provide*, or *supply*. Many passive functives can be readily converted into active ones simply by

Some Acceptable Verbs:

Absorb	Control	Hide	Minimize	Rotate
Actuate	Convert	Hold	Modulate	Satisfy
Aid	Create	Ignite	Mount	Seal
Allow	Direct	Impart	Move	Secure
Amplify	Ease	Impede	Open	Shield
Apply	Emit	Induce	Position	Shorten
Assist	Emphasize	Inject	Preserve	Space
Assure	Enclose	Instruct	Prevent	Standardize
Avoid	Ensure	Insulate	Promulgate	Steer
Change	Establish	Interrupt	Protect	Support
Close	Exude	Limit	Receive	Suspend
Collect	Facilitate	Locate	Rectify	Time
Comfort	Fasten	Maintain	Reduce	Tolerate
Conduct	Filter	Maximize	Repel	Transfer
Contain	Guard	Mesh	Resist	Transmit

Some Acceptable Nouns:

Access	Decoration	Flux	Noise	Task
Aesthetics	Density	Force	Odor	Time
Area	Dependability	Friction	Oxidation	Torque
Care	Deterioration	Heat	Pressure	Uniformity
Catalysis	Direction	Horsepower	Protection	User
Chromaticity	Dust	Image	Radiation	Variation
Color	Emissivity	Information	Repair	Vibration
Corrosion	Energy	Injury	Rust	Voltage
Current	Flow	Insulation	Stability	Volume
Damage	Fluid	Light	Status	Weight

Figure 6–4. A functive is a combination of a verb, preferably demonstrable on a nonverbal level, and a noun, preferably something measurable.

changing the noun to a verb. For example, "provide cooling" can become "cool space."

Maintain a constant frame of reference. Define functions performed specifically by the product under study. Do not confuse the viewpoint of the FAST diagram by shifting to functions performed by the user, or to those performed by the designer or manufacturer.

Avoid using the name of a part, labor operation, or activity as the noun. This weakens the function viewpoint of the diagram. In addition, it tends to limit later creativity by presuming the need for that part, labor operation, or activity.

A useful aid in choosing just the right word—both nouns and verbs—is *Roget's Thesaurus*. It is now also available as a computer program or in a hand-held display unit. A dictionary is also a useful reference work, but should not be regarded as a constraint; it is not necessary to restrict oneself to dictionary words if all members of the team thoroughly understand an "invented" word. Such a word is often better than the more prosaic standard words of the English language, since the members identify with it emotionally. This leads to a greater identification with the problem and tends to foster a broader creativity.

When preparing a FAST diagram, first post the rules and procedure, to guide the team through the creation of the diagram. There are four general rules:

1. Two words only: one verb, one noun.

2. Avoid the verbs "be" or "provide."

3. Noun is not a part, activity, or operation.

4. Maintain the viewpoint of the user.

The procedure consists of eight steps:

1. Define 40 to 60 functions.

2. Choose tentative task function.

3. Divide into basic and supporting functions.

4. Identify primary basic functions.

5. Identify primary supporting functions. Divide into four groups.

6. Expand diagram to right. (Reminder: Always branch.)

7. Rethink task. Verify diagram.

8. Add lines and numbers.

Step 1: Function Definition

Define 40 to 60 functions performed by the product or by any of its parts or labor operations (elements or activities of a process, service, or procedure). Simply ask, repeatedly, "What does (it) do?" Substitute for (it), in turn, every element, feature, labor operation, activity, material element, specification, tolerance, or requirement. In short, define the functions of every element of the product or process that costs money.

Record each function on a separate card (Fig. 6–5); cards with an adhesive strip on the back (such as the 3M Post-it ™ Note Pad; cut down to 1½ by 2½ inches) are best, because that simplifies display and allows for rearrangement.

If the product has been segmented so that two or more teams are able to share the study load, each team will initially prepare its own FAST diagram. The versions will be blended into one optimum diagram after Step 7.

Place the cards in a group in the lower right of a flip chart (Fig. 6–6).

Step 2: Select the Task Function

When 40 to 60 functives have been defined, start the FAST diagram by choosing the one functive from the flip chart that appears to be the task function, and placing it on the left of the chart (Fig. 6–7). The task function is the one that fulfills the overall needs and wants of the user—in other words, is the main reason for the existence of the product or process in the eyes of the customer or user.

It is often difficult for a team to agree on the task function this early. The initial selection of task function is therefore tentative. For example, the task function of a vacuum cleaner is not simply "what it does" ("suck air"); it is "what the customer desires" ("remove debris"). This tentative selection of task function is reconsidered after Step 3.

Step 3: Group into Basic and Supporting Functions

Move basic functions to the top right of the flip chart, leaving supporting functions in the lower right. To do this, simply ask, "Is the function essential to the performance of

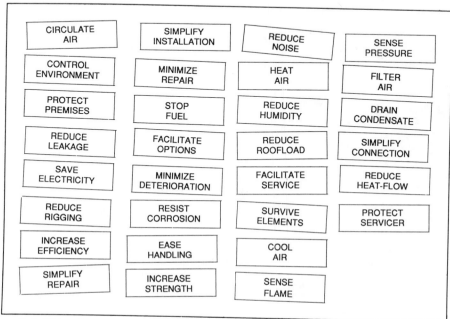

Figure 6–5. Thirty functives are recorded on cards displayed on the flip chart.

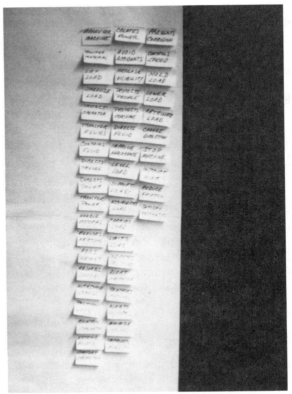

Figure 6–6. Cards recording all the functions are grouped in the lower right of the flip chart.

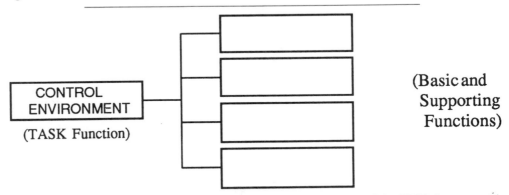

Figure 6–7. This block diagram illustrates the fundamental structure of the FAST diagram, with the task function moved to the left side of the flip chart.

the task function?" If the answer is yes, the function is basic . If the answer is no, the function is a supporting one. In Figure 6–8, this division has been done for functions describing the "Theme Thread" heat pump.

Basic functions, therefore, are essential to the performance of the task function. They fulfill the basic needs of the user.

Supporting functions are not essential to the performance of the task function, but

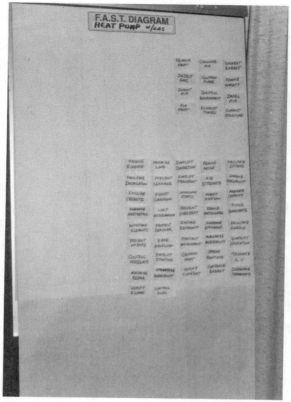

Figure 6–8. *This FAST diagram for the "Theme Thread" heat pump shows its basic functions grouped at the top and the supporting functions grouped at the bottom.*

are essential to increase product acceptance by satisfying the wants of the user. They equate to the quality of the product.

Supporting functions are the primary focus of user-oriented value analysis.

Supporting functions are essential to a successful product.

They are the basis for the user's perception of how well the product performs the task function.

They invariably form the basis for the user's buying decision.

Being related to the user's desires, they are primarily intangible, subjective, and therefore often overlooked in product design.

They distinguish a product from its competitors.

They commonly account for 75 percent of the cost.

The stylized coffeepot shown in Figure 6–9 illustrates the essence of the function classification concept. Clearly, the task function of the complete coffeepot (the two-word function that fulfills the overall needs and wants of the user) is "brew coffee."

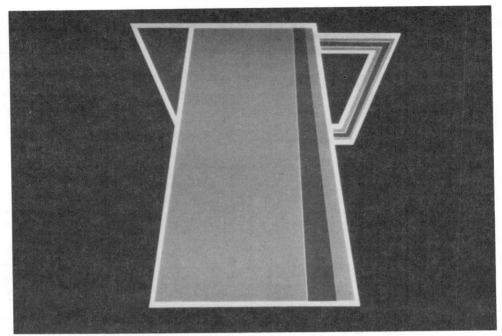

Figure 6–9. This stylized coffeepot with spout and handle illustrates the essence of classifying functions.

The function of the spout alone (Fig. 6–10) is probably "direct coffee." The question to be resolved in order to classify this function as basic or supporting is simply "Is the function 'direct coffee' essential to the performance of the task function, 'brew coffee'?" The answer is clearly no. "Direct coffee" is a supporting function.

The function of the basic container, excluding spout and handle (Fig. 6–11), is probably to "contain liquid." The same question about whether this function is essential is answered yes. Clearly one cannot "brew coffee" unless one can "contain liquid." This function is basic.

The function of the handle portion of the coffeepot (Fig. 6–12) is variously defined as "guide pouring," "maintain control," or, more commonly, "prevent burns." The essentialness question here is answered no. In order to "brew coffee"—the task function—it is not essential that we "prevent burns." As with the spout, it is a supporting function.

Step 4: Identify Primary Basic Functions

From the basic functions at the top right of the flip chart, select the ones that directly answer the question "How does it perform the task function?" The answer must be complete within itself. If a so-called conditioning phrase must be used to bridge the gap, look for a new answer.

If all of the direct answers to this question are not already on the flip chart, write a new functive on a card and place it on the diagram.

If a basic function does not directly answer the question "How does it perform the

Figure 6–10. This version of the coffeepot has only a spout attached; the function of that part is a supporting one.

task function?" leave it at the top right of the flip chart. It will be added to the diagram during Step 6.

Primary basic functions are placed at the top of the first column to the right of the task function (Fig. 6–13).

Step 5: Identify Primary Supporting Functions

All FAST diagrams contain the same four primary supporting functions. These have stood the test of time over nearly three decades. Grouping functions under the general headings of dependability, convenience, enhancement, and attractiveness is a powerful dynamic in focusing the problem-solving effort chiefly because these four areas make up all of the elements of that elusive product attribute: quality. Here are some examples of these elements, which are grouped under the four primary supporting functions:

Assure dependability: those functions that tend to minimize deterioration (related to added strength, corrosion protection, protection of people or the environment, protection of the product, elements that improve reliability of operation, and so on).

Assure convenience: those functions that make the product easier or more convenient to use (instructions, aids to servicing, cleaning, repairing, correcting, spatial relationships, ergonomics, and so on).

Enhance product: those functions that raise the product above customary expectations (modifications such as smaller, faster, lighter, added features, increased physical comfort, status, desire fulfillment, and so on).

Figure 6—11. The container portion of the coffeepot has a function that is basic—essential to the task function.

Please senses: those functions that appeal to the senses, both physical and aesthetic (such as appearance, noise level, and implications of performance, sturdiness, speed, and so on).

Write the four primary supporting functions on cards and place them in the first column to the right of the task function, below the primary basic functions (Fig. 6—14).

Step 6: Expand Diagram to Right

Next, expand each basic and supporting function to the right, asking "How does (the product) do this?"

Most of the answers will be found in the functions already filled out (and located) on the flip chart. Add functions as needed. Constantly maintain the viewpoint of the user throughout this questioning. Perform the expansion first on the basic function structure (Fig. 6—15), then on the supporting function structure.

Much thought and discussion are involved in creating a logically valid tree-structured diagram that directly complies with the "how" rule and involves no bridging or conditioning phrases.

As an example of the bridging or conditioning phrase that should be avoided, consider two functions of an internal combustion engine. The team defined function "convert energy" as a primary basic function. In expanding the diagram to the right, the team asked, "How does the engine convert energy?" One of the answers was, "By generating spark." The team reasoned that you need a spark to ignite the gasoline/air vapor in

Figure 6–12. The handle portion of the coffeepot has a supporting function, just as the spout does.

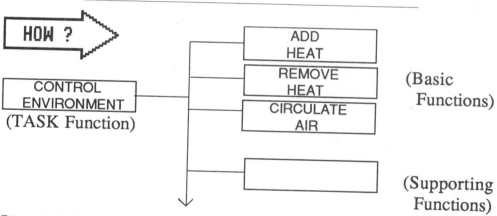

Figure 6–13. Figure 6–7 is expanded, with primary basic functions filled in and an arrow marked "How?" added.

order to convert energy. Such a conditioning or bridging phrase implies either that the logic is invalid or, as in this case, that another function belongs between the other two —in this case, "ignite fuel" (Fig. 6–16).

Expansion of the four primary supporting functions to the right is best performed in two steps. First, divide up all supporting functions into the four categories, by moving each of the cards to the right of the primary supporting function to which it relates. Next, assemble a tree structure for each of the primary supporting functions. This is the

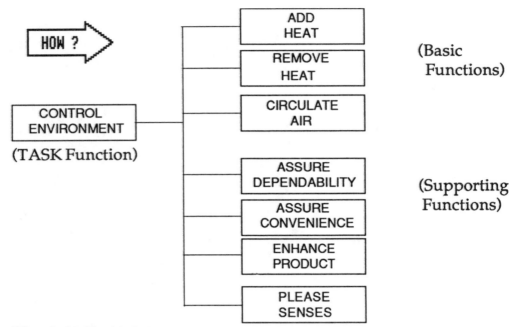

Figure 6–14. The block diagram is expanded further, with the addition of the four standard primary supporting functions.

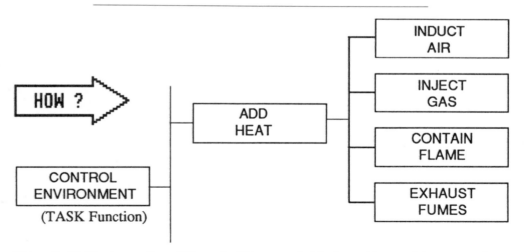

Figure 6–15. The top portion of Figure 6–13 is expanded further, with one of the primary basic functions now including four others to the right, answering a "how" question.

final grouping activity in the FAST process. This is done by looking over the supporting functions and determining whether there are two or more that fit into logical groups.

In Figure 6–17, three of the dependability functions, "resist corrosion," "reduce stress," and "minimize wear," can be logically grouped under an added function: "extend life." Simply write that function on a card, place it on the diagram to the right of "assure dependability," and then place the three answers to "How does the product

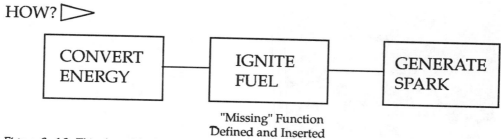

"Missing" Function
Defined and Inserted

Figure 6–16. This three-block diagram shows the identification and insertion of a missing function, filling in a gap.

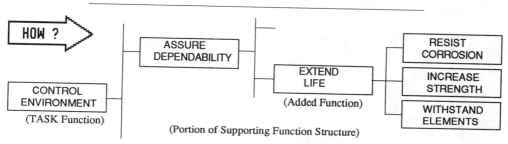

Figure 6–17. One of the primary supporting functions is expanded using the grouping technique, with an added function inserted in between.

extend life?" to the right of it. Continue grouping the supporting functions, adding new ones whenever appropriate.

Expansion of the diagram to the right should stop when the answer to the "how" question requires using a part name, labor operation, or activity.

This branching seen in the examples always results from such "how" questioning. When the branching stops, the diagram is complete. It is pointless to continue expansion after branching is complete; a nonbranching network is simply a string of redefinitions of a single function, serving no useful purpose.

The diagram now displays all of the causes and all of the consequences of each functional requirement.

Step 7: Verify the Diagram

The question "How . . . ?" is the inverse of the question "Why . . . ?"; consequence is the inverse of cause. The final step in the preparation of a FAST diagram is to verify that this inverse rule has not been violated (Fig. 6–18).

Start at the right end of each branch of the diagram. Ask, "Why does the product or process perform this function?" If the function to the left of it in the diagram directly answers that question, with no bridging or conditioning phrases, the diagram is valid. If not, correct the how/why logic of the diagram.

Melding the Diagrams

If the product being FAST-diagrammed has been segmented so that two or more teams are sharing the study load, each team at this point in the process will have prepared

Figure 6–18. A member of a value analysis team verifies the validity of the "how" structure by applying the "why" rule, doublechecking each branch from right to left this time.

somewhat different versions of the FAST diagram. These must be melded into one diagram, which will serve all of the teams for the remainder of the study.

A formal melding session is essential, since each of the teams has invariably developed a sense of ownership and a very real conviction of the superiority of its version, through the unconscious process known as cognitive dissonance reduction. This is an unconscious mental process that always makes public decisions more apparently correct in the mind of the decision maker.

The melding session should be guided by the value analysis coordinator and led by the project engineer or other person who is fully familiar with the technology of the product (Fig. 6–19). With the support of all of the members of the teams studying the product, that person will create a common FAST diagram by extracting the most appropriate function cards from each of the other diagrams. This tends to be an emotional give-and-take session, and requires of the leader something close to the wisdom of Solomon and the compassion of a clergyman. A major benefit of such a melding session is a deeper and broader confirmation of the teams' understanding of the product and its relationship to user/customers.

At this point, the team enters on its completed FAST diagram a series of lines defining the branches. The team also numbers each of the functions in an alphanumeric hierarchy. The completed FAST diagram of a heat pump with auxiliary gas heating is shown in Figure 6–20.

Figure 6–19. A project engineer leads a multiteam session to meld two FAST diagrams into one. The diagrams are posted side by side on the wall.

The FAST diagram is now ready for the final step: the translation of our customary item-cost viewpoint into the less confining framework of function cost.

Loading the Unconscious Mind

The FAST diagram is a semantic analog of the product under study. In Figure 6–20, the 46 independent functions (those at the ends of the branches) are totally equivalent to the 4-ton heat pump. The diagram thus constitutes a new viewpoint to focus the creative problem-solving effort.

The vigorous mental effort involved in converting the product into its semantic equivalent has a second purpose, more fundamental and more powerful than the obvious objective of shifting the problem solver's view from product to function. This purpose is to load the collective minds of the team with a greater quantity of valid data about the product than any single mind has held before; all of this data is expressed in the unconstraining language of function—a common language, created by the team itself.

FUNCTION-COST

The objective of this effort is to establish the cost of each of the functions noted on the FAST diagram. The process comprises three steps:

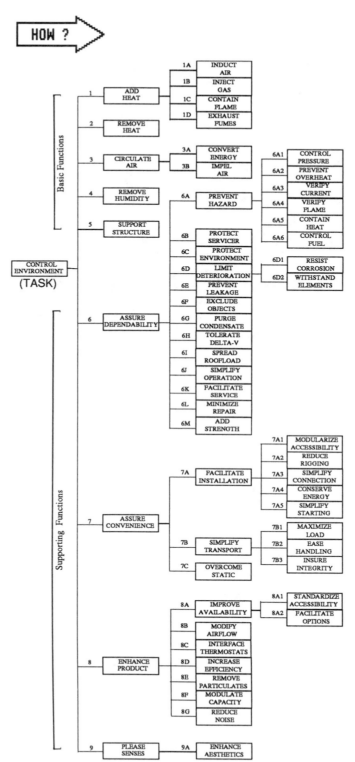

Figure 6–20. The FAST diagram for the "Theme Thread" product—the heat pump—is completed, including branch lines and a hierarchy of numbers for the functions.

1. Define: Divide the total cost of the product or process into its smallest elements.

2. Allocate: Determine what function is performed by each of those elements.

3. Post: Add together all of the costs that perform each function.

The process is applied to the present design or the base case of the product under study.

The greatest power of value analysis is realized when a function-cost structure is also prepared for each significant competing product.

Define Function-Costs

In a manufactured product, cost is commonly defined in a document called a costed bill of materials and labor, abbreviated CBOML. In its most common form, the indented CBOML (see Table 4–2), it is the ideal structure for function-cost allocation. The ICBOML is ideal for cost breakdown in preparation of an effective costed FAST diagram. The example in Table 4–2 describes a manufactured product. A similar ICBOML can be prepared for a process, service, or procedure simply by restructuring the available data into this format.

Allocate

Simply ask, of each independent line in the ICBOML, "What does it do?"

This question often has more than one answer. Each one must be one of the functions on the FAST diagram. If the function does not exist on the diagram, add it.

Only the functions at the far right of the diagram qualify for allocation. Any function that is not at the right end of a branch of the FAST diagram is not an independent function; it is merely the name of a so-called summing point for the branches to its right.

Single Allocation

If the line item performs a single function, enter the number of that function on that line.

An illustration of single-function allocation is shown in Figure 6–21. This is a simple and obvious allocation; all of the labor and material perform the same function. Many allocations are not so straightforward. Often the same line item of cost performs several functions.

SINGLE FUNCTION ALLOCATION

(Blower Motor)	MATERIAL	LABOR	TOTAL	FUNCTION
End Bell - Operation 050, Prime & Paint	**$.0211**	**$.0590**	**$.0801**	**6D1**

Q: What does the labor and material of operation 050, "Prime & Paint," do?

A: It protects the steel from rusting. It therefore performs Function # 6D1, "**Resist Corrosion.**"

Figure 6–21. This tabular presentation is an example of single-function allocation of cost.

Multiple Allocation

It is often necessary to split a single line item of cost among several functions. The logic for such splitting is governed by three rules:

1. Opinions are not allowed. For instance, if it is determined that a line item of cost performs two functions, it is not permissible simply to allocate half of the cost to each.

2. Each split must have a measurable and auditable basis.

3. There are two permissible methods of splitting: particle analysis and equivalent allocation.

In an example of splitting, the team member in Figure 6–22 is indicating a function cost of $0. In this value analysis of an oil field valve, the team members are splitting the $147 cost of a wedge into the functions it is performing. They first determined that the item performs 14 functions. Upon review, however, they decided that 3 of these were not really performed at all, and that 6 were free, in that they were automatically performed when another, more essential function was accomplished. This left 5 functions, which were then costed by particle analysis or equivalent allocation.

Figure 6–22. A team member sits at a flip chart and fills in a worksheet on function-cost splitting of an oil field valve part.

Particle Analysis

The most common technique of splitting is to allocate elements of labor or material to the function that they perform. An illustration of one of the ways this is used to split the costs of a solid piece of aluminum is illustrated in Figure 6–23.

A team split the function costs of the aluminum extrusion, part of a ceiling-mounted air-control louver. The members determined that various portions, or so-called particles of the aluminum extrusion performed a total of five different functions.

The portion marked "A" performed the function "prevent leakage," since it held a flexible strip to seal each louver against the next one.

The portion marked "B" was for mounting on the shaft around which it rotated. The function was defined as "control air."

The portion marked "C" is a 90-degree element that strengthens the louver, but primarily it ensures that each louver will present a straight surface to guarantee sealing against the flexible strip in the adjacent one. It thus also serves to "prevent leakage."

The portion marked "E" is the minimal thickness that can perform the "control air" function. The portion marked "D" is the extra thickness to ensure that the louver will not deform under extreme conditions of operation; this extra thickness performs the function "withstand abuse."

When allocating function costs, treat each element or particle of material, or each time element or particle of labor, separately. Determine how much cost is added to the product or process to provide that particle's function, then allocate that cost to that function.

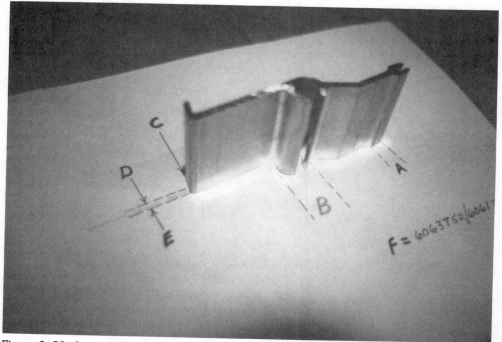

Figure 6–23. A section of aluminum extrusion for an air control device is divided into particles, indicated with letters, for function-cost allocation.

An illustration of particle analysis allocation is shown in Figure 6–24. In this example, involving particles of labor, it was determined that holding the tight tolerances required 0.05 hours—one-third—of the 0.15 total hours of machining labor. Therefore, $.2115 (.05 hours at $4.23 per hour) was allocated to "increase efficiency," function 7D on the FAST diagram.

The balance of the labor cost ($.4230) represents the shop tolerance labor. It performs the basic function "circulate air" by simply holding the armature in place in the stator and permitting it to do its job.

Equivalent Allocation

Where particle analysis is not applicable, creative effort must be applied to determine an equivalent to some portion of the line item of cost. That portion of cost is allocated to the function performed by that equivalent, with the remainder allocated to the other function(s) performed by the line item.

Ask, "What is the simplest way I can perform the prime reason for the existence of this item or operation?" The cost of that equivalent is allocated to that prime reason function.

When choosing an equivalent, avoid a pendulum swing to the ridiculous. When allocating a military door latch, for example, do not choose a screen door hook and eye as an equivalent. Whatever is selected as an equivalent must be a reasonable and practical alternative to the present method of performing the prime reason function (Fig. 6–25).

Figure 6–26 illustrates an actual example of equivalent allocation that a team at Dominion Bridge of Calgary, Ontario, performed. The company was the major Canadian producer of oil well pump jacks, and it had assigned two powerful, multinational value analysis teams to perform an unrestricted VA-based redesign. In their analysis of the drive system, the teams approached the function-cost splitting of the $840 herringbone drive gear shown in Figure 6–27.

If the gear reduction system were not required to perform function 5A2B, "reduce wear," a pair of helical gears costing only $380 would be completely sufficient to perform the prime function, 1C2, "lift pump." The helical gears are thus an equivalent, used to split the $840 total cost into its two components.

LABOR SPLIT

(Blower Motor)	LABOR HRS	LABOR $	TOTAL	FUNCTION-COST
End Bell, Op 010, Machine Bore	.1500	$.6345	$.6345	7D $.2115 (Tolerance)
				4 $.4230 (Balance)

Q: What does the labor operation do that machines a bearing bore in the housing?

A: It creates the bearing bore that holds the shaft of the Armature in place. The Armature couples with the stator to generate the Starter's torque. The bore is machined to a dimensional tolerance of + or - .001" and a concentricity of .002" TIR. This is to maintain the bearing precisely centered in the Starter Motor, permitting a smaller magnetic air gap, making the Motor more efficient.

Figure 6–24. This tabular presentation is an example of function-cost allocation by the particle analysis technique.

Figure 6–25. A cutter blade element of an agricultural machine chopper system, left, is shown with a serrated kitchen knife, which performs a function equivalent to the blade's prime reason function.

PURCHASED MATERIAL SPLIT

(Oil Well Pump-Jack Transmission)	MATERIAL COST	FUNCTION-COST
Herringbone Main Drive Gear	*$840*	1C2 **LIFT PUMP** $380
		5A2B **Reduce Wear** $460 (Balance)

Q: What does the Herringbone Gear do?

A: It is the main drive element that withstands the whole load of the pumping structure, as well as the load of the entire pump-shaft string, and the load of a tall column of oil. Its *Prime Function* is LIFT PUMP, Function 1C2.

An EQUIVALENT that performs only the *Prime Function* was identified. It was determined that a pair of helical gears at $190 each ($380 total) would perform the *Prime Function* alone. This was therefore used as an EQUIVALENT to permit the splitting of the cost of the Herringbone Gear.

It was determined that the helical gears would create a huge side load on the gearshaft bearings. The Herringbone Gear therefore effectively adds a feature to the EQUIVALENT item. That feature is the *reduction of side load on the bearings*. Since side load would greatly increase wear on the mechanism, the Function added through the use of a Herribone Gear is *REDUCE WEAR*, Function number 5A2B.

Figure 6–26. This tabular presentation is an example of function-cost allocation by the equivalent allocation method.

Figure 6–27. Gearbox, with cover removed, showing top of the herringbone main drive gear.

Note that the term *balance* in the examples shown in Figures 6–24 and 6–26 refers to the cost that is left over after the initial cost-splitting activity. This balance is always allocated to the function that describes the prime reason for the existence of the item being analyzed.

Testing Split Allocation Validity

Apply the following test to each function-cost split that anyone suspects might be based on opinion instead of measurable and verifiable fact: If the costs associated with one of the functions were eliminated, would the remaining costs permit the other function(s) to be performed?

If the answer to this question is no, the split is not valid.

In the examples shown in Figures 6–24 and 6–26, the answer is clearly yes. If, for instance, the tolerance cost shown in Figure 6–24 were removed, the loose-toleranced machine bore operation would, indeed, completely perform the "circulate air" function. The only function lost is the "increase efficiency" function, which results from the tighter tolerance.

As an example of an invalid function-cost split, visualize a single bolt that holds in place two structures. It is not correct to split the cost 50–50 between the functions performed by each structure. With the bolt present, both functions are being performed. If 50 percent of the bolt were removed, neither function would be performed.

When a split fails the validity question, allocate the total cost element to the function that is judged to be the fundamental reason for the existence of the cost, and assign zero cost to the other function(s). In effect, the other function gets a free ride.

Allocation of Purchased Parts

Resist the temptation to shortcut the process when allocating the cost of a purchased part. As shown in the foregoing examples, most parts perform several functions. When the total cost of a purchased part is allocated to a single function, it seriously distorts the function-cost structure of the complete product.

The general rule on purchased parts is to create an estimated bill of materials and labor, including each element of raw material and each labor operation. Each is then allocated to the functions that it performs.

If sufficient talent is not available within the team to perform this bill-of-materials structuring with even approximate accuracy, outside support should be sought.

An alternate method of splitting the cost of a purchased part is to estimate the increase in price for each part feature. In one instance, a drip-proof electric motor cost $834; an open-frame motor was available for $680. The $154 difference clearly performs the function "resist moisture."

Post Function-Cost Allocations

As each of the function-cost allocations is completed, the team recorder enters the information onto a function-cost worksheet. When all allocations have been recorded, each column of the worksheet is totaled (Fig. 6–28), and the resulting total function-costs are posted on the FAST diagram. After all individual function costs have been posted and totaled, enter the total function-costs in a separate column on the FAST diagram (Fig. 6–29).

Each member of a value analysis team that has just completed function-cost allocation is better informed on the microanatomy of the design and its costs and its functions than any other single person has ever been; and their collective understanding is in terms of their own common language—that of the FAST diagram. This massive but rational collection of data is a prime source of the power of the process.

FUNCTION-ACCEPTANCE

While we now know what each function costs, we do not yet know what each function is worth. The final step in the information phase of the job plan is to establish, for each function, a measure of its worth.

Webster's Ninth New Collegiate Dictionary defines worth in part as "the value of something measured by its qualities or by the esteem in which it is held." This is not very measurable by usual standards, is it? And it has been said that if you cannot measure it, you probably don't understand it.

In value analysis, we define cost and worth as follows: Cost is defined as the sum of labor and material to produce a product and/or to maintain a product over its useful life cycle. Worth is defined as the degree to which an item fulfills the user's needs and wants.

These are entirely separate matters. While both are ultimately measured in dollars, the dollar measure of worth is extraordinarily difficult to pinpoint. In fact, the worth of any item is not a fixed figure; it varies with changes in the user's needs and desires.

How, then, can we define this will-o'-the-wisp? We have stated that value is worth divided by cost. It is clear that unless we can define worth, we cannot define value. The

Function Number / Part or Operation	Cost	1B INJECT GAS	2 REMOVE HEAT	3B IMPEL AIR	6A2 PREVENT OVERHEAT	6A5 CONTAIN HEAT	6A6 CONTROL FUEL	6B PROTECT SERVICER	6D1 RESIST CORROSION	6E PREVENT LEAKAGE	6H TOLERATE DELTA-V	6K FACILITATE SERVICE	6L MINIMIZE REPAIR	6M ADD STRENGTH	7B2 EASE HANDLING	7B3 INSURE INTEGRITY	8B MODIFY AIRFLOW	8C INTERFACE THERMO.	8D INCREASE EFFICIENCY	9A ENHANCE AESTHETICS
Panel Liner - Steel Mat'l	4.66					4.66														
Op 010 Notch	.44					.44														
Op 020 Pierce	.54											.54								
Op 030 Form	1.20					1.20														
Op 040 Wash	.15								.15											
Op 050 Prime/Paint	.56								.56											
Control Module (Purch)	34.20										12.10							22.10		
Access Panel (Purch)	9.12							2.22				6.90								
Gas Valve (2) (Purch)	64.22	32.11					32.11													
Compressor Timer (Purch)	31.40				31.40															
V-Belt, Heavy Duty (Purch)	13.80			9.60									4.20							
Condenser Tube Mat'l	71.20		53.60											17.60						
Op 010 Form Ends	2.21									2.21										
Condenser Fins Mat'l	14.50																		14.50	
Op 010 Cutoff	2.41																		2.41	
Op 020 Form	8.15																		8.15	
Op 030 Pierce, Extrude	5.14																		5.14	
Op 040 Lance Louvers	5.42																5.42			
Cond.Ass'y - expand tubes	11.55																11.55			
Lifting Brackets - St. Mat'l	7.20														7.20					
Op 010 Cutoff	.65														.65					
Op 020 Punch (6)	1.02											.68				.34				
Op 030 Form	.94														.94					
Op 040 Wash	.15																			.15
LiftingBrkt Gusset St Mat'l	6.99													6.99						
Op 010 Blank	.70													.70						
Op 020 Wash	.15																			.15
Lftg Bkt Assy Op 010-Weld	.75													.75						
Op 020 Prime/Paint	.30																			.30
Bolts, 1/4-20 X 1.5 (24)	2.64											2.64								
Nuts, 1/4-20 (24)	1.92											1.92								
L'washers, 1/4 (24)	1.20											1.20								
TOTAL COSTS	305.48	32.11	53.60	9.60	31.40	6.30	32.11	2.22	.71	2.21	12.10	13.88	4.20	26.04	8.79	.34	16.97	22.10	30.20	.60

Figure 6–28. This is a portion of a filled-in function-cost worksheet on the "Theme Thread" heat pump.

answer to this seeming paradox is really quite simple: Since the user defines worth, simply ask the user.

A system has evolved from the basic work of Thomas J. Snodgrass that effectively measures the relative worth of the functions on the FAST diagram. This system is based on: (1) the collection of data reflecting the user/customer attitude toward the product, and (2) the allocation of that data (in precisely the same manner as the cost data was allocated) to the FAST diagram.

Once the user/customer data have been allocated to the FAST diagram, an approximation of the worth of a function can be established by evaluating the clustering of that data around each function. This does not result in a purely numerical rating of function worth, but it does permit an informed, if subjective, judgment of relative worth. In the analysis phase, the teams will balance this judgment against the numerical cost of each function in order to identify areas of opportunities for value improvement.

The Measurement of User/Customer Acceptance

Prior to the start of the team effort, data are collected by either the questionnaire method or the focus panel method (see chapter 5). The data from either of these methods are similar, and the same techniques of allocation are used.

With the former method, the interviewer's raw questionnaires are annotated in a joint

Figure 6–29. *A team member enters the allocated function costs into the right-hand column of a FAST diagram.*

session by the value analysis coordinator and marketing before being turned over to the teams for data allocation. This annotation consists of identifying and circling, classifying and rating each unique allocable element of each response (see Fig. 5–3).

The raw data collected at the focus panel session are analyzed by a computer program, and a report is prepared listing each like and dislike with the associated mode and range of its vote (see Fig. 5–6). This includes competitive data where applicable.

Data Allocation

Acting as discussion groups, the teams then determine, for each of the elements of the questionnaire data or each like or dislike of the focus panel data, the answer to the following question: "To what function on the FAST diagram was the respondent referring?"

The functions must be those on the FAST diagram. Only the task function and the four primary supporting functions are ineligible for allocation, since they are really only category headings. New functions may not be added at this stage, to avoid distortion of the previously established function-cost structure.

Each team member notes the number of the function on the questionnaires or focus panel report. If the team determines that the respondent was truly referring to more than one function, all applicable function numbers should be noted. Normally, however, a response contains only one key thought.

When all allocations are complete, the team posts the data in the right-hand column of the FAST diagram, opposite the function to which the response was allocated. Any allocation to other than a "right-most" function is posted adjacent to that function.

Each entry has two parts: The first is the mode of each focus panel vote or the rating of each questionnaire element. The second is the number of the like or dislike or the questionnaire question number. The entry for like number 13 with a vote mode of 8, for instance, would be "8–13." These numbers are included to make the data easier to trace, simplifying later evaluation of it in identifying value analysis targets.

Use an appropriately colored marker for each entry on the FAST diagram; green might be used for likes or pluses and red for dislikes or minuses. For later printed reports showing this data, preface the dislikes or minuses with the letter *d*.

A completely allocated FAST diagram of the heat pump in tabular form is shown in Figure 6–30. This method of allocating user attitudes directly onto the FAST diagram was first used by Thomas F. Cook.

COMBINEX© TO DEFINE WORTH?

The seeming complexity of both the questionnaire and focus panel methods for establishing function worth has led some to experiment with a simpler decision matrix method, such as the Combinex© method (see chapter 10).

Such experiments have failed, due to the highly structured nature of the Combinex© process. In order to measure the true worth of a function, the user/customer data must be completely unstructured and voluntary—a stream-of-consciousness representing the actual, real-time reactions of users/customers to a live situation. The decision matrix cannot establish the worth of functions. Although it is highly effective in rating alternatives on the basis of rigidly defined weighted criteria, any data obtained through Combinex© that purport to relate to function worth are very likely to mislead the team members.

SUMMARY OF THE FAST PROCESS

The FAST diagram is now complete. It represents a precise semantic equivalent of the project under study. Each function has been annotated with data reflecting its function-cost and its function acceptance. We are ready for the analysis phase, in which we will identify the value analysis targets.

The greatest power of the FAST diagram is not, however, as a deathless document for future reference. Its primary benefit is in the mental struggling and interaction of team members in developing the diagram. This forced introspection generates a deep and broad understanding of the way the product does its job. It loads the minds of the participants with a huge quantity of valid data, all expressed in the unconstraining

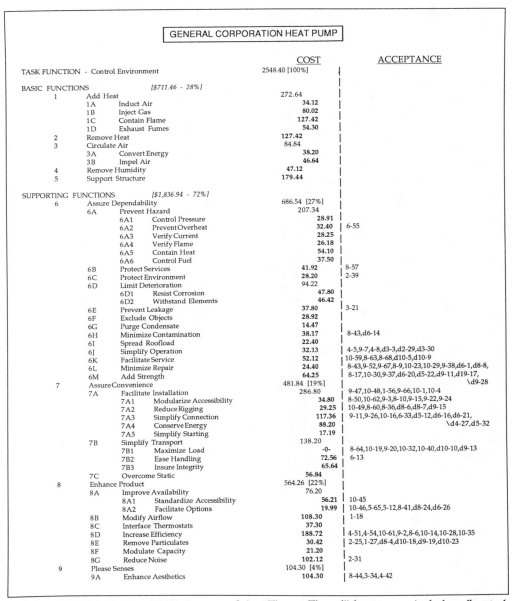

GENERAL CORPORATION HEAT PUMP

			COST	ACCEPTANCE
TASK FUNCTION - Control Environment			2548.40 [100%]	
BASIC FUNCTIONS		[$711.46 - 28%]		
1	Add Heat		272.64	
	1A	Induct Air	34.12	
	1B	Inject Gas	80.02	
	1C	Contain Flame	127.42	
	1D	Exhaust Fumes	54.30	
2	Remove Heat		127.42	
3	Circulate Air		84.84	
	3A	Convert Energy	38.20	
	3B	Impel Air	46.64	
4	Remove Humidity		47.12	
5	Support Structure		179.44	
SUPPORTING FUNCTIONS		[$1,836.94 - 72%]		
6	Assure Dependability		686.54 [27%]	
	6A	Prevent Hazard	207.34	
		6A1 Control Pressure	28.91	
		6A2 Prevent Overheat	32.40	6-55
		6A3 Verify Current	28.25	
		6A4 Verify Flame	26.18	
		6A5 Contain Heat	54.10	
		6A6 Control Fuel	37.50	
	6B	Protect Services	41.92	8-57
	6C	Protect Environment	28.20	2-39
	6D	Limit Deterioration	94.22	
		6D1 Resist Corrosion	47.80	
		6D2 Withstand Elements	46.42	
	6E	Prevent Leakage	37.80	3-21
	6F	Exclude Objects	28.92	
	6G	Purge Condensate	14.47	
	6H	Minimize Contamination	38.17	8-43,d6-14
	6I	Spread Roofload	22.40	
	6J	Simplify Operation	32.13	4-5,9-7,4-8,d3-3,d2-29,d3-30
	6K	Facilitate Service	52.12	10-59,8-63,8-68,d10-5,d10-9
	6L	Minimize Repair	24.40	8-43,9-52,9-67,8-9,10-23,10-29,9-38,d6-1,d8-8,
	6M	Add Strength	64.25	8-17,10-30,9-37,d6-20,d5-22,d9-11,d19-17,
7	Assure Convenience		481.84 [19%]	\d9-28
	7A	Facilitate Installation	286.80	9-47,10-48,1-56,9-66,10-1,10-4
		7A1 Modularize Accessibility	34.80	8-50,10-62,9-3,8-10,9-15,9-22,9-24
		7A2 Reduce Rigging	29.25	10-49,8-60,8-36,d8-6,d8-7,d9-15
		7A3 Simplify Connection	117.36	9-11,9-26,10-16,6-33,d5-12,d6-16,d6-21,
		7A4 Conserve Energy	88.20	\d4-27,d5-32
		7A5 Simplify Starting	17.19	
	7B	Simplify Transport	138.20	
		7B1 Maximize Load	-0-	8-64,10-19,9-20,10-32,10-40,d10-10,d9-13
		7B2 Ease Handling	72.56	6-13
		7B3 Insure Integrity	65.64	
	7C	Overcome Static	56.84	
8	Enhance Product		564.26 [22%]	
	8A	Improve Availability	76.20	
		8A1 Standardize Accessibility	56.21	10-45
		8A2 Facilitate Options	19.99	10-46,5-65,5-12,8-41,d8-24,d6-26
	8B	Modify Airflow	108.30	1-18
	8C	Interface Thermostats	37.30	
	8D	Increase Efficiency	188.72	4-51,4-54,10-61,9-2,8-6,10-14,10-28,10-35
	8E	Remove Particulates	30.42	2-25,1-27,d8-4,d10-18,d9-19,d10-23
	8F	Modulate Capacity	21.20	
	8G	Reduce Noise	102.12	2-31
9	Please Senses		104.30 [4%]	
	9A	Enhance Aesthetics	104.30	8-44,3-34,4-42

Figure 6–30. A complete FAST diagram of the "Theme Thread" heat pump includes allocated costs and allocated user/customer acceptance data. Dollar figures in **bold** are allocated function costs; all other figures are sums of function costs to their right in the tree. As is typical, no respondents reacted in any way to basic functions.

language of function. This mentally stored data is the raw material for the creative and synthesis phases to follow.

The intense microallocation of costs and user attitudes is the first step of the optimum creative problem-solving process: "Load the mind with pertinent information and identify the value problems" (see Fig. 3–4). Following closely upon the equally intense mental exercise involved in generating the FAST diagram, it serves to lock into the minds of the team members the reasons for the very existence of each and every element of the product or process. This develops an unprecedented breadth and depth of understanding of the product and its users and customers.

The microallocation technique is a substitute for the clutter that characterized some early FAST diagrams, which tended to include functions that described actions, using nouns that were parts or labor operations. Inclusion of such hardware-oriented functions did, indeed, deepen the team's understanding of the product under study, but also tended to clutter the FAST diagram with unnecessary complexity. Microallocation produces an even deeper understanding of the product while maintaining the FAST diagram as the analog of the essence of the product objectives. Its primary goal is to shift the team's viewpoint toward user-defined objectives and away from the physical product. Inclusion in the FAST diagram of hardware-oriented, labor-oriented, or activity-oriented functions detracts from that goal.

THE COMPARATIVE VALUE ANALYSIS SYSTEM

As discussed in chapter 2, comparative value analysis through use of a FAST diagram has the unique capacity to compare the functions of each element of the product directly on a one-to-one basis, with any other product that accomplishes the same purpose. The detailed procedure of the identification and analysis processes are described below, illustrated with an actual example of a 2000cc automobile engine.

Preparing the Comparative FAST Diagram

The first step is to identify competitive products or processes. For value analysis purposes, it is unnecessary to evaluate more than two competitors. Both should be in the same class as the product being analyzed, in that they should be models that compete directly in the market. They should generally comply with the description of the "high side" (Fig. 6–31) and "maverick" (Fig. 6–32) as discussed in chapter 5.

The value analysis team then creates a FAST diagram describing the product under study. It next verifies that the FAST diagram also describes the competing products.

The cost-allocation procedures described earlier are then applied to each of the products in turn: first the product under study, then those of each competitor.

The costs of the product under study are available in an indented, costed bill of materials and labor, the labor presented in the form of costed detail routings. The costs of each of the competitive products or processes must also be made available in similar format and detail. This requirement commonly terrifies general management. It is our experience, however, that a competent industrial engineer, given access to data, can complete such a document in a few days for each competing product. As described in chapter 4 (see page 49), the procedure consists of preparing a set of numbered sketches or marked-up drawings and then estimating the cost of each, with support from industrial engineering and purchasing.

The attitude collection procedures described in chapter 5 are then applied to each of the three competing products in turn. This effort may involve formal questionnaires and face-to-face or telephone interviews, or it may, where appropriate, take the form of a user/customer focus panel.

Figure 6–31. The photo shows a "high side" competing product, specifically a Lincoln Town Car.

It is important that the questionnaires or user focus panels for the competing products be conducted anonymously.

The user attitude data then is posted onto the comparative FAST diagram, using a summary notation. This briefer format minimizes the excessive clutter that would result from entering each response individually with its associated notations for simplified tracing later.

The procedure is simply to sum the ratings of all of the likes identified in the questionnaire or focus panel. Enter that sum, preceded by a plus sign, in the appropriate column. Repeat the process for dislikes, using a minus sign.

A completed summary sheet illustrating the comparative value analysis system is shown in Figure 6–33. It contains an unprecedented volume of cost and worth data. The team's approach to the analysis of this data will vary greatly with the character and quantity of the acceptance data.

The Value Standard

Note that the right-hand column of Figure 6–33 represents the *value standard,* a term that Thomas J. Snodgrass adapted from the early value standards concept of Roy Fountain, the key collaborator with Lawrence D. Miles on the development of the value analysis system. Snodgrass's value standard was used as the ultimate aiming point, a maximum cost-reduction target that could not be surpassed. Indeed, it was a target that, for all practical purposes, could not even be reached.

Compare, for example, the function costs for function 2, "resolve forces," in Figure

Figure 6–32. *This is an example of a "maverick" competing product, specifically a Lamborghini Countach.*

6–33. The client's engine accomplishes this for $156.65. The "high side" competitor spends $190.82. The value standard cost for this function is $153.37, which is less than any of the other three. How can this be? The reason becomes clear when we analyze the source of each of the four figures:

For function 2A, the lowest cost is the client's, at $64.18.

For function 2B, the lowest cost is the "maverick's," at $84.84.

For function 2C, the lowest cost is the client's, at $4.35.

The total function cost of function 2 is the sum of these three, or $153.37, a figure that is lower than the "resolve forces" cost of any of the three competitors individually.

When following this selection and summing procedure for all of the function costs, we arrive at a value standard for the total engine of $804.12, which is 30 percent below the present product cost and 33 percent below the present cost of the "high side" as well as 12 percent below the cost of the "maverick" product.

This 30 percent figure is a rough measure of the cost-reduction potential of the comparative value analysis system.

The value standard concept is based on a fundamental assumption that even a weak competitor has optimized certain of the elements of its design.

Interface Costs

It is clear that such an artificially calculated figure as the $153.37 value standard for the function "resolve forces" is an impossible target. Within the constraints of presently applied technology, it can never be reached, only approached. The reason is that each of the elements making it up is from a totally different product, with each product often based upon a different design concept. In most cases, these elements cannot just be plugged in to an ideal value standard design without the addition of some so-called interface costs.

The value standard, therefore, must be regarded as merely an artifice, and not a complete and practical alternative design. Invariably, however, it aims the problem-solving process in potentially fruitful directions.

[AUTOMOBILE ENGINE]	HIGH SIDE		CLIENT		MAVERICK		VALUE STANDARD
TASK FUNCTION:	COST		COST		COST		COST
Deliver Torque	1,204.12		1,145.65		910.24		804.12
	Cost	ACCEPTANCE	Cost	ACCEPTANCE	Cost	ACCEPTANCE	Cost
BASIC FUNCTIONS:							
1 Convert Energy	356.60		368.04		356.20		334.92
1A Induct Air	33.27		34.12		30.34		30.34
1B Inject Fuel	77.74		80.02		82.68		75.68
1B1 Pump Fuel	19.46		17.40		18.28		17.40
1B2 Meter Fuel	58.28		62.62		64.40		58.28
1C Distribute Vapor	17.20		12.44		14.12		12.44
1D Control Flow	49.92		54.30		48.64	-6	48.30
1D1 Start Flow	17.80		16.18		16.50		16.18
1D2 Stop Flow	32.12		38.12		32.14		32.12
1E Ignite Fuel	179.47		187.16		180.42		168.16
1E1 Generate Spark	51.24		62.60		61.24		51.24
1E2 Distribute Spark	83.80		84.30		78.52		78.52
1E3 Deliver Spark	10.10		6.12		8.46		6.12
1E4 Time Spark	33.33	-8	34.14		32.20		32.20
2 Resolve Forces	190.82		156.65		158.54		153.37
2A Contain Explosion	92.24		64.18		69.20		64.18
2B Transmit Force	92.18		88.12		84.84		84.84
2C Transmit Torque	6.40		4.35		4.50		4.35
3 Couple Torque	58.12		84.84		76.90		58.12
4 Exhaust Fumes	112.28	-19	109.41	-18	112.05		102.96
4A Direct Exhaust	61.44		52.12		54.41		52.12
4B Control Flow	50.84		57.29		57.64		50.84
4B1 Start Flow	16.43		19.13		18.46		16.43
4B2 Stop Flow	34.41		38.16		39.18		34.41
SUPPORTING FUNCTIONS:							
5 Assure Dependability	800.43		775.05		532.71		499.12
5A Maintain Integrity	370.05		333.11		185.84		179.69
5A1 Minimize Leakage	135.40	+24	108.12	+15 -3	42.20	+43 -4	42.20
5A2 Minimize Loosening	46.25	+10 -5	62.13	+12 -38	51.82	+24 -12	46.25
5A3 Resist Breakage	188.40	+18 -6	162.86	+10 -9	91.24	+28 -14	91.24
5B Survive Environment	52.18		71.92		66.02		46.24
5B1 Clean Air	6.60	+34 -4	8.90		6.80		6.60
5B2 Exclude Water	22.24		38.67		41.82	+32 -62	22.24
5B3 Absorb Shock	23.30	+75	24.35	+48 -30	17.40	+30 -42	17.40
5C Enhance Life	378.20		370.02		280.85		273.19
5C1 Reduce Stress	137.61		109.36	-62	81.64	+42 -38	81.64
5C1A Increase Modulus	112.34	+10 -96	87.12		60.64		60.64
5C1B Increase Inertia	25.27		22.24		21.00		21.00
5C2 Reduce Corrosion	49.26		41.13		26.29		26.29
5C3 Reduce Friction	110.62	+55 -12	153.56		110.81		108.61
5C3A Reduce Contact	42.20	+112	76.34	+13 -37	44.40	+18 -18	42.20
5C3B Lube Surfaces	68.42	+52 -92	77.22	+48 -60	66.41	+38 -18	66.41
5C4 Remove Contaminants	10.12	+17 -8	9.20	+92	8.55	+80 -8	8.55
5C5 Minimize Imbalance	38.41		18.78	+78 -13	26.44	+21	18.78
5C6 Remove Heat	32.18		37.99	+23 -77	27.12	+34 -11	27.12

Figure 6–33. This portion of a completely cost- and acceptance-allocated FAST diagram allows comparison of the internal combustion engines made by the client and two competitors. Forty-four of the 62 functions are shown.

Acceptance Data

The value standard column in Figure 6–32 is strictly cost-based. No attempt has yet been made to integrate the function costs with the user acceptance data. Indeed, for example, note the $17.40 figure chosen as the value standard for function 5B3, "absorb shock." Although it is the lowest cost (the others are $23.30 and $24.35), it does not represent an acceptable alternative for the user/customer. In the acceptance column for the $17.40 cost of the "maverick's" configuration are the notations " + 30" and " − 42"; the " + 30" means that the users/customers regard this function as rather important. The " − 42" means that they are unhappy with the method used by the "maverick" to accomplish this function.

Now the team must decide whether the value standard of $17.40 is valid and, if it is not, which of the other two function costs should be entered in its place. In this example, the team first reviewed the user data to determine the source of each of the acceptance indices, then arrived at the following conclusions:

"Maverick." The "dislikes" or faults/complaints of the users that resulted in the acceptance index of − 42 are nontolerated.

Client. The − 30 index also represents nontolerated dislikes.

"High side." With no negative index whatever, coupled with the + 75 high index for "likes," and with a cost 4 percent less than even the client cost for this function, this is clearly the value standard for this function. The cost-based value standard of $17.40 is therefore replaced with the worth-based figure of $23.30.

This process is repeated for each independent function of the allocated comparative FAST diagram.

THE IDEA BANK

Throughout the information phase, the processes of cost allocation and attitude allocation involve dynamic interaction of team members, with regular excursions into a creative mode of behavior. During this period, it is common for solutions to various problems to occur to a team member. It is important to capture these creative thoughts so they may be fed later into the problem-solving process during the synthesis phase. The mechanism provided for this is simply a blank notebook, kept in an accessible location and titled the Idea Bank.

As an idea arises, the team member is encouraged to jot down a one-line summary in the Idea Bank, adding his or her initials. This captures the thought for later use and permits team members to return, without fear of losing an idea, from their temporary creative excursion to the more prosaic analytic mode of the information phase.

Chapter 7

THE ANALYSIS PHASE

FOCUSING CREATIVITY (A PARADOX?)

The mechanism of the analysis phase uses the value analysis target identification process. It has six steps, leading up to the focusing of creativity on targeted functions.

The creative mind works best in the absence of constraints. This is either expressed or implied by all writers on the creative process. "From time to time, we must turn off our judicial mind and light up our creative mind. And we must wait long enough before turning up our judicial light again. Otherwise, premature judgment may douse our creative flames, and even wash away ideas already generated." (Osborn 1953, 95). The principle of suppression of judgment during the idea-generating process is a fundamental precept in the process of brainstorming, developed by Alex F. Osborn, Ph.M. In emphasizing the essential nature of this freedom-from-judgment principle, Osborn said, "We sometimes have to keep our minds open by shutting out environment" (Osborn 1953, 128).

How in a value analysis study, can we justify focusing our creativity? How can we concentrate our problem-solving effort toward the specific areas where creative solutions are required? Doesn't it follow from Osborn's caveat that when we focus our creative effort we are "turning on our judicial mind," thereby dousing our creative flames?

In modern value analysis, we do focus our creativity, and this act of focusing does not shut out environment. It does not compromise our creativity.

The six steps of the value analysis targeting process are as follows:

1. Define and analyze functions.

2. Allocate costs to functions.

3. Allocate user/customer attitudes to functions.

4. Compare function-costs to function-attitudes.

5. Identify value analysis targets.

6. Focus creativity on targeted functions.

Steps one through three are explained in detail in chapter 6. Steps four and five constitute the analysis phase of the value analysis job plan and are described in detail in this chapter. Step six is covered in chapter 8.

The objective of the analysis phase is to identify where the value analysis team should concentrate its creative effort for maximum impact.

If the team were to use the customary shotgun approach to problem solving, it would systematically attack all 40 to 60 functions on the FAST diagram. The value analysis targeting process focuses the problem-solving effort on only a few of these functions. This well-aimed rifle-shot approach assumes that most of the functions do not represent problems that need to be solved. It concentrates the attention of the team only on those functions where there is a poor match between function-cost and function-acceptance.

It is futile, after all, to attempt to solve problems where none exist.

IDENTIFYING VALUE ANALYSIS TARGETS

The value analysis targeting procedure is performed immediately following the information phase. Then there are several days of incubation before the creative phase.

A value analysis target exists whenever the function-cost and function-acceptance factors are out of balance.

This can be seen in the completed FAST diagram of the "Theme Thread" heat pump (see Fig. 6–29), which provides a graphic display of the value of each function: It displays both the total and individual function costs; it also displays the acceptance factors for each function.

There are 27 possible different combinations of the cost and acceptance factors shown below for each function on the diagram. The challenge of the analysis phase is to determine which of the functions on the FAST diagram represent the best opportunities for improvement.

FUNCTION COST	FUNCTION IMPORTANCE	FUNCTION FAULTS
High	High	Nontolerated
Low	Low	Tolerated
None	None	None

Value Analysis Targeting Procedure

The team members must assume a new form of behavior in order to identify value analysis targets. Up to this point, they have been functioning in an objective mode. Now they must function in a totally subjective, customer-focused mode.

The team gathers informally around the allocated FAST diagram in a discussion group. Each function on the diagram is separately considered as a candidate for value analysis target.

The team will consider, at least, the following:

- function-cost
- function-attitude (user/customer feedback)
- team member knowledge
- implementation planning checklist
- preconceived solutions

When a function is defined as a candidate for value analysis target status, it is recorded on a flip chart, together with its function cost (Fig. 7–1).

If, after all the functions have been discussed, the flip chart contains more than about eight, the team should review its selections and attempt to reduce the number to eight, or ten at the very most.

This list of functions comprises the targets, or focal points for the creative phase. The total of these targets commonly make up only about 15 to 20 percent of the total

Figure 7–1. *A team member lists potential value analysis targets on a flip chart.*

number of functions on the FAST diagram, though they often constitute 60 to 85 percent of the total cost.

The crop dryer in Figure 7–2 illustrates a classic value analysis target. The "ease repair" function costs only 1.4 percent of the total, but members of the user focus panel indicated that the function is very important to them. In addition, the users identified several significant faults in the present design.

The team focused its problem-solving effort on this function and implemented changes in the design that corrected the user faults and also reduced the cost of the function from 1.4 percent to 1.25 percent of the total.

Value analysis targeting satisfies a key objective of the team: to identify value analysis projects. Doing so in terms of functions, not hardware, is the prime secret of the success of value analysis. Focusing creativity in this unconstrained manner avoids the Alex Osborn "turning up our judicial light" constraint. Indeed, it enhances the creative effort.

WHERE IS "MY" FUNCTION?

Most team members arrive with a private agenda, a list of changes they would like to see implemented. Some resist the focusing procedure of the value analysis target method if their function does not survive on the final list of candidates; that is, if, in their

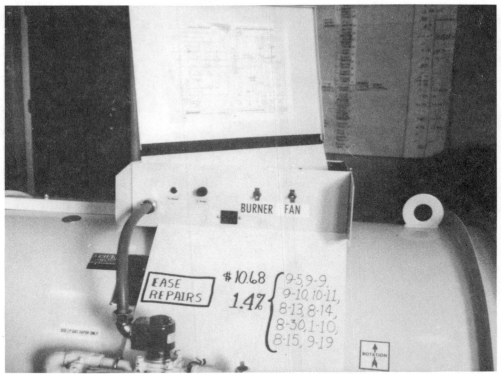

Figure 7–2. The card attached to a crop dryer indicates cost and attitude data for one function, which involves a small percentage of total costs but is considered important as a value analysis target.

opinion, the limited list of targeted functions does not include the ones they feel will trigger consideration of their design changes.

This is a misplaced concern. The list of targeted functions does not narrow the viewpoint of the problem solvers. It does not wipe from their minds all previously stored information that is unrelated to these few functions. It merely focuses the creative behavior of the team toward a selected few of the functions. The private agenda of each team member will still emerge in the highly unconstrained environment of the creativity phase.

Figure 7–3 is a portion of the FAST diagram reflecting the $28 million annual cost of overhead and selling, general and administrative expenses of a major division of an international corporation. The right-hand column lists the costs and percentages of each function. Notations reflecting function worth are also entered, adjacent to the appropriate functions.

The team, comprising the president and his staff, reviewed this data and concluded that the outlined functions sould serve as value analysis targets. In each targeted function, the function cost, the function importance, and the function faults were, in the team's opinion, not balanced. The outlined functions served to focus their later efforts in the creative phase.

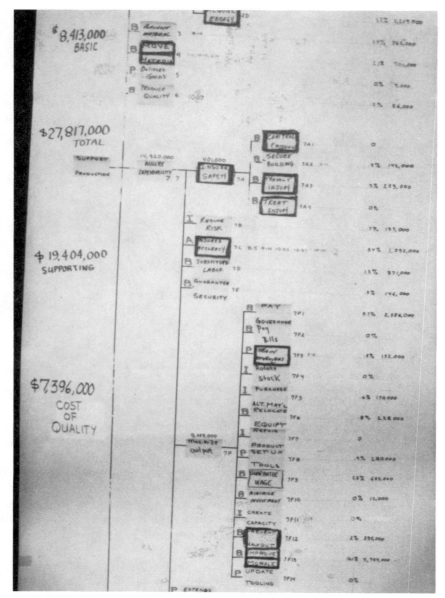

Figure 7–3. FAST diagram outlining costs, functions, and value analysis targets.

INTERSESSION EFFORT

Team members are instructed not to think consciously about their value analysis project during the two- to five-day break between sessions. They are told that following this instruction will permit the data to "incubate," letting their unconscious minds generate solutions.

This incubation is step two of the optimum creative problem-solving process: "Divert the mind to permit the unconscious mind to solve the problem" (see Fig. 3–4). The

Ideas, like eggs, need time to hatch.

collective minds of the team members have been loaded with pertinent data in step one, the information phase. These minds then have been permitted to sift and combine this data unconsciously while the team members were otherwise occupied. The next step in the value analysis system is to create a stimulating environment. The team is ready for the creativity phase.

Chapter 8

THE CREATIVITY PHASE

CREATIVITY OR INNOVATION?

We need creativity! We have heard this plea many times in many places, but is that really what we need? Perhaps what we need is innovation; that is, applied, implemented creativity—in other words, results.

Theodore Levitt, editor of the *Harvard Business Review,* had some provocative comments on this subject several years ago. He said that advocates of creativity

> have generally failed to distinguish between the relatively easy process of being creative in the abstract and the infinitely more difficult process of being innovationist in the concrete.
>
> Having a new idea can be "creative" in the abstract, but destructive in actual operation, often hindering a company instead of helping it.
>
> Suppose you had two artists. One tells you an idea for a great painting, but he doesn't paint it. The other has the same idea, but paints it. You could easily say the second man is a great creative artist. But could you easily say the same thing for the first man? Obviously not. He is a talker, not a painter.
>
> That is precisely the problem with so much of today's pithy praise of creativity in business . . . with the unending flow of speeches, books, articles and "creativity workshops" whose purpose is to produce more imaginative and creative managers and companies. They mistake brilliant talk for constructive action.
>
> . . . A powerful new idea can kick around unused in a company for years, not because its merits are not recognized, but because nobody has assumed the responsibility for converting it from words into action. What is often lacking is not creativity in the idea-creating sense, but innovation in the action-producing sense, i.e., putting ideas to work.[1]

The focus of modern value analysis is on putting ideas to work. This is not a creativity workshop where we simply let our minds roam. We focus our creative talents only

---

[1] "Creativity Is Not Enough," by Theodore Levitt, May–June, 1963." Copyright © 1963 by the President and Fellows of Harvard College; all rights reserved. Reprinted by permission of *Harvard Business Review.*

on areas that three days of intense effort have proven are in need of creative improvement.

Value analysis is not creativity for its own sake. Each team is an innovative design team, working toward a clearly established goal. Value analysis is results-oriented.

CREATE BY FUNCTION

In the creativity phase session, the focus is on the functive. A flip chart is headed with one of the functions identified in the analysis phase as a value analysis target. This serves two objectives: First, it maintains the critical perspective of function, the "what it does" for the customer, rather than the "what it is" in the mechanical or procedural sense. Second, it displays ideas, words, concepts, and solutions on a flip chart. This fulfills an essential dynamic of the brainstorming process: the immediate visual feedback of team output, both to the individual who expressed the word or concept and to the problem-focused team.

BRAINSTORMING

The ideal creative problem-solving method is called brainstorming. This method of generating creative solutions was first published in 1953 by Dr. Alex Osborn (*Applied Imagination,* New York: Charles Scribner's Sons), as a method of generating new ideas on how to advertise products. Early attempts to apply brainstorming to the problems of business were often ineffectual, due to the absence of systems to elevate the flights of fancy to the realities of the industrial world. But brainstorming is the ideal creative process for value analysis. Its unique structure takes effective advantage of the function orientation of the value analysis system. The method of elevation described in chapter 9 provides the needed bridge to business realities.

There are dozens of formalized systems of creative problem solving. While some of them—for instance, the Crawford Technique or Synectics®—are very effective in and of themselves, they are unable to take advantage of the power of the functional approach and they do not directly treat the denominator of the value equation: cost.

To provide a receptive atmosphere for the germination of creative ideas, the brainstorming process is controlled by these four rules:

1. No criticism. Judgment is temporarily suspended. One cannot allow oneself to be critical while at the same time being creative. As Alex Osborn said, "If you try to get hot and cold water out of the same faucet at the same time, you will get only tepid water. And if you try to criticize *and* create at the same time, you can't turn on either *cold* enough criticism or *hot* enough ideas" (Osborn 1953, 301).

2. Free wheel. The wilder the better. The techniques of chapter 9 will later tame this wildness.

3. Go for quantity. Research has proven that, as more thoughts are generated, the quality of each individual thought improves. Quantity, therefore, begets quality.

4. Combine and improve. Chain together the thoughts already displayed, and expand upon them.

These four rules, in combination with the abstract nature of the function written at the top of the flip chart, result in a broad and uninhibited variety of ideas. Some of these entries under the heading may be concepts, or even solutions; most, however, will be no more than words or phrases—at best, verbalizations of hazily formed mental images (Fig. 8–1).

In the 40 years since Dr. Osborn invented brainstorming, one problem has repeatedly surfaced: Ideas seem to come in bursts. When a burst peters out, it is difficult to tell whether the team has reached a plateau—that is, whether it has truly tapered off—or has merely paused to collect itself. Modern value analysis has developed a procedure to make this determination. It is termed the *plateau breaker.*

Give each team member the goal of writing 5 new thoughts on a piece of paper within three minutes. This commonly results in 20 or 30 new thoughts, proving to the team members that they hadn't really plateaued at all.

Get Off Route 25

Charles F. Kettering, director of research for the General Motors Corporation, had a lifelong passion for improving the creativity of his engineers.

At the age of 80, in an address to the Dayton, Ohio, Education Association, he repeated his famous Route 25 story. Here is a paraphrase:

I live in Dayton, but I can't get a job here, so I work in Detroit. I drive back and forth on weekends and it takes me about 4½ hours.

Figure 8–1. A flip-chart shows 50 thoughts recorded during brainstorming about one of the functions of the "Theme Thread" heat pump.

An associate told me that I couldn't possibly get to Dayton from Detroit in under 5 hours.
I invited him to ride with me that weekend, and we made it in 4½ hours.

He said, "But you didn't stay on Route 25!"

I spend a good part of my time telling my young engineers to get off route 25.

Wild Card

An experience in 1966 led the author to experiment with over 100 value analysis teams
using a creative problem-solving variation called "wild card." The team is asked, during
its brainstorming, to pick out the most useless entry on the flip chart. The leader then
accepts the challenge of guiding the team in the elevation of that ridiculous word or
phrase into a novel, useful, and implementable solution within three minutes.

The original objective of the experiment was to prove to team members that they
should cast off all inhibitions and record even the ridiculous. It invariably performs this
motivational function, and incidentally it provides the team with a pattern to follow in
the synthesis phase, in which the objective will be to elevate the flip-chart chaff into
usable ideas and concepts.

An example of this three-minute wild card technique is shown in Figure 8–2.

Solo Brainstorming?

Brainstorming is clearly a group activity. Two of its four rules apply only to group idea
generation. It is dramatically effective when value analysis teams use it in generating
hundreds of useful words, ideas, and concepts in a few concentrated hours. It is also,
however, amazingly effective when carefully used by an individual.

The plateau breaker technique described above is a powerful recharger of the team's
creative battery precisely because it satisfies the ownership needs of the individual team
member. In three minutes, each individual team member invariably generates five or
more new ideas. This burst of creativity is released when the constraints of the team
environment are lifted—when the member starts practicing, for a brief moment, solo
brainstorming.

Dr. Ernest Benger, head of research at Du Pont, supported the notion of solo brainstorming in a note to his far-flung staff: "No idea has ever been generated except in a
single human mind . . ." (Osborn 1953, 289).

Those who practice brainstorming outside the team environment tell us that they
require a form of pump priming. They find that they are most effective after they have
participated in group brainstorming, which in effect sets the pattern for uninhibited idea
generation. After such mind-setting, they find that solo brainstorming can add dramatically to their store of ideas, concepts, and solutions.

When to Stop Brainstorming

It is something of a judgment call, but generally speaking, saturation is reached within
20 to 40 minutes of brainstorming per function. The goal is to record 50 to 100 thoughts
on the flip chart for each function.

To aid the team in the next phase, display each flip-chart sheet on the wall as it is
filled (Fig. 8–3).

Wild Card Creativity

PRODUCT: AIR MOVER, Aerovent, Inc.
TARGETED FUNCTION: "Relieve Stress"

This propellor assembly is used to move large volumes of air. It comprises a hub with several cast aluminum blades. Each blade is heat treated and then reshaped after the flash is ground off. The FAST Diagram Function "Relieve Stress" described these activities. Its cost was an inordinate percentage of the total, so it was identified as a Targeted Function. The left flip-chart below displays five of the words/phrases that were written by the team members. They were asked to identify the most ridiculous of the entries and they chose "Hit Frank," an entry which resulted from the failure of the Value Analysis Coordinator, Frank Swisshelm, to provide an acceptable luncheon earlier in the day. They felt that to "Hit Frank" was a way to "Relieve Stress". The elevation effort led to an aluminum baseball bat, a concept which was elevated in under three minutes to the novel and practical solution to their blade problem shown in the right flip-chart below.

FUNCTION: "RELIEVE STRESS"
WORDS / IDEAS:
> HEAT TREAT
> FORGE
> DON'T GRIND
> VALIUM
> HIT FRANK

WORD: "HIT FRANK"
ELEVATION:
> RELIEVE STRESS
> BASEBALL BAT
> HICKORY
> HANK AARON
> ALUMINUM

WORD: "ALUMINUM BAT"
FORCE FIT:

> MAKE THE FAN BLADES FROM THIN-WALL ALUMINUM TUBING. TAPER WITH HYDROFORMING AND FORM INTO BLADES USING MATCHED DIES. ATTACH A CAST BASE END AND A CAST TIP WITH MAGNEFORMING ®.

The team members agreed that the drastic change from a sand-cast blade to one made from tubing would never have been considered without the phrase "Hit Frank" and the force-fitting process. They also saw the new blade concept as having many ancillary advantages, including a smooth surface for airflow and the elimination of a balancing operation, since the new concept is inherently balanced.

Figure 8–2. *The process of elevating the ridiculous phrase "hit Frank" to the practical tubing-based blade is an example of the wild card technique.*

Environment

The brainstorming session is the third step of the optimum creative problem-solving process: "Create an environment that stimulates the unconscious mind to deliver its solutions" (see Fig. 3–4).

Many ideas, concepts, and solutions will follow as a result of the value analysis creativity phase. A large proportion of these will be consciously drawn from the notes

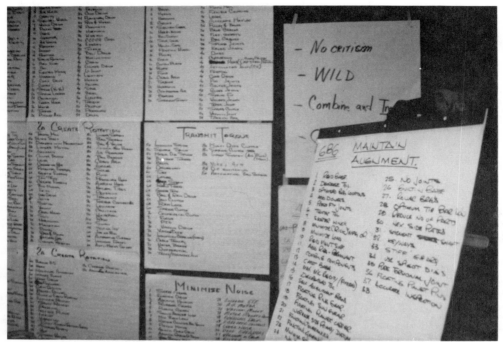

Figure 8–3. A wall is covered with flip-chart sheets full of words and ideas born during brain-storming in the creativity phase.

and the memories of team members and their associates. A significant minority of these will be unique: They are the output of the team's collective unconscious minds and represent the highest order of accomplishment of the value analysis system.

Chapter 9

THE SYNTHESIS PHASE

THE CHALLENGE

The creativity phase has generated a large quantity of flip-chart entries. It is common for a team to develop 400 to 500 entries, triggered by the eight or so value analysis targeted functions.

Buried within these hundreds of entries are the seeds of unique solutions. The objective of the synthesis phase is to force each entry to reveal these seeds, and then to build, expand, and flesh out each of them until at least one member of the team recognizes it as a potential solution.

The synthesis phase is actually a several-week period during which each champion of a potential solution investigates the feasibility of the concepts that he or she has volunteered to shepherd through the process. The objective of the synthesis session is to kick off the phase, briefly define the concepts, and outline a plan for investigating their feasibility.

ELEVATION OF WORDS TO SOLUTIONS

The output of a brainstorming session is a large quantity of generated thoughts. These should not be dignified by being called ideas. They are no more than a series of mental images, expressed as words or phrases and entered on a flip chart. In an effective brainstorming session, remember, judgment is temporarily suspended to ensure uninhibited purging of the collective minds of the team. Each uncriticized thought, therefore, represents an unconscious mental image. The limitations of language and the often cryptic nature of the method by which the unconscious mind communicates with the outside world prevents these images from being translated in a form that is directly usable in a rational environment.

Since problems are solved creatively only in the unconscious mind, it is crucial to accept that each entry on the flip chart originated as a mental image. Every attempt must be made to translate, or elevate, each entry into a form that can be put to use by either the individual who created it or another member of the value analysis team.

Simply by repeatedly asking "How can we use it?" the team establishes a nurturing environment for a not-yet-born idea. The effect is to draw out and translate into words the image in the mind of the original speaker, and to draw out as well the mental images triggered in the minds of the other team members.

Avoid Trying to "Leap" Too Far

Elevating a flip-chart entry from word to idea, and finally to concept and solution, is not a simple, one-step process. In the words of Britain's former prime minister, David Lloyd George, "The most difficult thing in the world to do is to leap a chasm in more than one jump."

The chasm between word and solution is too great to leap successfully all at once.

By definition, the word is no more than the verbalization of a mental image. The image tends to fade away and die if we manhandle it by trying to force the word to reveal its hidden usefulness. It is necessary that we place some stepping-stones, so to speak, between the word and the solution; then, carefully maintaining a positive view-point, we can attempt to elevate each word to an idea, next elevate each surviving idea to a concept, and finally elevate each surviving concept to a solution.

Unlike the word, an idea has at least the essence of form; it can be related to and evaluated. A concept has form. It can be sketched and measured and its cost can be estimated. Feasibility can be tested. Finally, a solution has drawings and specifications and firm quotations and test results. It can be implemented.

Consider All Words and Ideas

In the words of prolific inventor Charles F. Kettering, "Man is so constituted as to reject a new thought because of its 10 percent wrongness . . ." It is important to resist this instinctive rejection of the words on a brainstorming list. Do not scratch out a word or later an idea simply because it does not appear to be a practical product improvement.

The principle upon which this rule is based is that if the word or idea looks foreign to the project under study, this simply means that it is unique. This is a valid reason to try to use it and certainly not a valid reason to scratch it!

An idea, being one of nature's most delicate creations, will not survive rough han-dling; but how much more delicate is a word, which is no more than a cryptic symbol of an image in the mind of the speaker!

GROUP WORDS TO SIMPLIFY ELEVATION

If the team has fully complied with the rules of brainstorming, the quantity of words on the flip charts could total 400 or more. This massive quantity of raw material, each item being no more than a verbalized mental image, is usually daunting to the team mem-bers. At first glance, an impossible task faces them. They are instructed to ask, of each item, "How can I use it?" They are further instructed not to treat any one of the items lightly. If they are disposed to calculate the enormity of the task before them, they might multiply 400 by 60 seconds and find, even if they limit the consideration of each word to one minute, that they are faced with a full day of rigorous elevation.

The team's apprehension is actually misplaced. Experience has proven that up to 1,000 words can be rigorously elevated in the six hours allocated to the task. To reduce the team's concern, however, a technique of idea grouping is initially applied to the words on the flip charts.

Procedure for Grouping

The team identifies flip-chart entries that appear to belong together. A letter or number is entered on the flip chart next to each member of that group. Further groups are identified until all entries have been grouped and annotated. It is common for each group to contain 5 to 15 entries. This reduces the challenge from 400 or so elevations to a manageable 100 or so. Perform the grouping operation on one flip-chart sheet at a time. Discontinue the grouping when the team agrees that the elevation task can be completed without further grouping.

The team then attempts to elevate each group of entries to the point where the members agree an idea or concept has arisen that should be considered for further investigation. During this elevation, the identification of each group itself no longer has significance; perform the elevation on the images generated by the specific entries, not on the arbitrarily chosen name of a group.

The grouping of flip-chart entries tends to smother the idea triggers that might be contained within each individual entry. To prevent this loss of raw material, quickly run through any remaining members of the group, asking of each entry, "Is there any other way I can use it?" If no further ideas or concepts arise, line out the entries (Fig. 9–1).

OWNERSHIP

When an idea or concept arises that is worthy of further investigation, it is recorded on the team master list synthesis phase worksheet. The team captain then asks the other members whether anyone is willing to be its champion. That person's initials are recorded on the team master list.

The team agrees on the key steps to be performed in the next intersession period in order to investigate and evaluate the concept. These are also recorded on the team master list, as is the team's opinion of the class of the concept, or ease of implementation (1 being the easiest, 3 the most difficult or elaborate). Figure 9–2 is an example of a completed team master list.

No idea, concept, or solution will be taken beyond this point unless a team member voluntarily accepts ownership. The member who agrees to champion the idea or concept enters it on his or her personal champion list synthesis phase worksheet, along with a more detailed description of the actions he or she will perform to test the concept (Fig. 9–3).

These actions generally include the following:

Verify the feasibility of the idea or concept by gathering data from experts within and outside the organization.

Estimate the economic consequences of the idea or concept, with regard to implementation cost as well as any possible cost reduction.

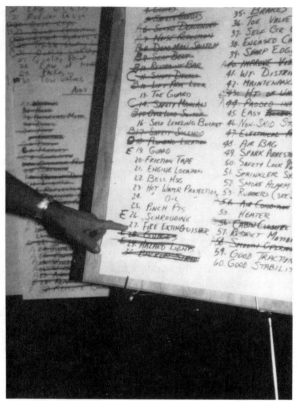

Figure 9–1. A team member examines a flip chart, on which some entries are annotated with letters and some have been lined out.

Estimate the effect of the idea or concept on user acceptance of the product.

Report this information back to the team at the development phase session.

All entries are eventually claimed by a champion or, if a diligent attempt at elevation fails, eliminated. This phase is complete when all entries are crossed out, for either reason.

The Champion's Approach

The mood of the champion should be similar to that of a research scientist. Rather than concentrating on what is wrong with an idea or concept, the approach should be to define the possible advantages; then try to prove that they are valid.

When a concept approached with this viewpoint has been proven worthy of further consideration, it can better withstand challenges. Only at that point is it appropriate to define the possible disadvantages; then modify and optimize the concept by overcoming them with additional ideas.

No.	IDEA OR CONCEPT	CHAMPION	PLAN FOR IMPLEMENTATION	CLASS
1	Reduce gauge from 24 to 22 on sides	GTW	Do Engineering Study. Verify Cost w/Purch.	2
2	Change to Spun Copper Dryer	FRK	Get Quotes. Verify Performance per ARI	2
3	Remove Elbow fm Compeland Comp	TCF	Draw new Configuration. Release to Purch.	1
4	Enhance Fins. Go to 12 FPI	FRK	Design & build five. Test per ARI	2
5	Use Auto-Strapping Machine	PMG	Perform Time Study on Auto-Strapping	1
6	Change OD Fan to Vertical	FRK	Build Lab Mockup. Test for Efficiency	3
7	Decrease Unit Width for Trucking	TOL	Perform Design Layout. Design Packaging	3
8	Single Point Power	FRK	Redesign Electrical Circuit and Cabling	3
9	Provide for Bottom Power Input	FMG	Design and Cost New Configuration	3
10	Remove Unused Process Tube	FRK	Calculate Cost Saving	1
11	Replace Sensor Support with Wire	TOL	Draw and Cost Wireform	1
12	Buy Wiring Harness Outside	FMG	Quote and Select Vendor	1
13	Deep Draw Evaporation Drain Pan	FRK	Draw and Cost New Pan	2
14	Change 4-Speed to 2-Speed Motor	FRK	Quote and Select Vendor. Run API Tests	2
15	Have Motor Vendor Integrate Brkt	FRK	Send Bracket to Vendor. Draw new Assy.	2
16	Move Low-V Block to Basic Unit	FMG	Redesign Harnesses and Mounting Plates	3
17	Replace Aluminum Blades w/CRES	FRK	Quote three sources	1
18	Change Copper Def Sensor to Plastic	TOL	Quote and Test Samples to ARI	2
19	Move Relay/Transformer to Base	BMD	Draw New Mount and Cabling Diagram	2
20	Change to Time-Temp Defrost	FRK	Design Control Unit. Estimate Cost	3
21	Install Programmable tester in Final	BMD	Cost the Three-Station System	2
22	Go to High EER Copeland Comp.	FRK	Cost and Run Efficiency Test	2

Figure 9–2. Part of the synthesis phase worksheet, this team master list is filled in with *22 ideas or concepts, including identification of champions, key investigative steps, and class, or ease of implementation.*

Why Insist on Ownership?

This ownership process does inevitably result in the loss of some potentially useful solutions. The most significant result, on the other hand, is that each championed idea or concept has the opportunity to be fully evaluated.

The power of the champion concept lies in the commitment that arises from ownership. Its secondary benefit is in the early-stage elimination of ideas and concepts that would have no chance to be implemented.

No.	IDEA OR CONCEPT	PLAN FOR IMPLEMENTATION	CLASS
2	Change to Spun Copper Dryer	Quote new Dryer. Set up five samples with Test Unit #1.	
		Arrange with Jackson to test with currect Evaporator Unit	2
4	Enhance Fins. Go to 12 FPI	Take Cartersville sample to Ken in Engineering. Have him	
		run a full Heat Dissipation Test. Run ARI Test	2
6	Change OD Fan to Vertical	Sketch new Config. Have Lab mockup and test for Efficiency	3
8	Sinagle Point Power	Contact Drafting re Laying Out. Cost Estimate new Cabling	3
10	Remove Unused Process Tube	Sketch non-Tube Compressor. Get quote from Copeland	1
13	Deep Draw Evaporator Drain Pan	Send Dwng to Ogarth Co. Get quote on Drawn Pan Version	2
14	Change 4-Speed to 2-Speed Motor	Get two quotes. Contact Parker for Labor Quote	2
15	Have Motor Vendor integrate Brkt	Sketch new assembly. Send bracket to Vendor for quote	2
17	Replace Aluminum Blades w/CRES	Get three quotes on Stainless blade version	1
20	Change to Time-Temp Defrost	Have Engng sketch new Control Unit. Cost with Purchasing	3
22	Got to High EER Coopeland Comp.	Get samples. Set up test with Lab. Cost change	2

Figure 9–3. The champion list of the synthesis phase worksheet is filled in with 11 concepts accepted by a single champion, and more detailed implementation plans.

On the surface, it would seem ill-advised to reject a concept that the team regarded as a great idea when it was first noted. In fact, however, if no champion can be found within the team that created the concept, it really wasn't a great idea at all, and it would never have been implemented.

Change is traumatic. Every person and every organization is afflicted, to a varying degree, with neophobia; the fear of something new. Change means risk, potential failure, and potential embarrassment. For a change to run this gauntlet successfully, it must have a champion, someone who voluntarily accepts ownership and nurses the concept through its development from a creativity phase idea to its acceptance as an implemented change. Without a champion, even the best idea or concept is doomed to failure.

As Peter Drucker summed it up, "Whenever *anything* is being accomplished, it is being done, I have learned, by a monomaniac with a mission" (Peters and Austin 1985, 135).

Withdraw from the Idea Bank

The entries that were deposited in the so-called idea bank notebook during the information phase (see chapter 6) are already ideas or concepts, rather than merely words. Review and flesh out each entry and enter it on the team master list. The team captain then asks for a volunteer champion. An idea book entry is often championed by the team member who initialed the original entry.

Review of Dislikes

When all possible ideas and concepts have been developed from the creativity phase flip-chart sheets, the team retrieves the user-attitude data sheets developed in the focus panel or through the questionnaires. The team reviews each dislike or fault to determine whether it is tolerated or nontolerated by the user. If the fault affects user acceptance of the product, it is nontolerated.

The team then reviews its list of ideas and concepts to make certain that all nontolerated dislikes are eliminated.

VENDORS AND EXPERTS

During the intersession period between the synthesis and development phases champions will be contacting vendors and experts, both inside and outside the organization. The pattern for this interaction is established in a formal, four-hour session during the last half day of the synthesis phase session.

One advantage of a value analysis system is that, in investigating their ideas or concepts, champions are forced to search outside their organizations. This is in recognition that, to varying degrees, all employees are provincial. Their primary viewpoint is within their organizations. The majority of the influences in our daily working life are internal influences.

In the early days of value analysis, Lawrence Miles urged value analysts to make use of vendors' engineering departments. Of the original Miles list of 13 techniques of value analysis, 3 focused on the advantages of going outside, looking beyond the limits of the organization.

In order to exemplify this interaction to the team members, to demonstrate the unlimited resources available to them, the vendor/expert session is scheduled at the end of the synthesis phase session.

The vendors and experts serve as sounding boards for the champions' ideas and concepts. It is common also for this four-hour interaction to result in a significant number of new proposals, triggered by suggestions from vendors or experts and then championed by team members.

Number of Vendors and Experts

The total number of vendors and experts should probably not exceed 20, representing as many as 12 organizations per product under study. This will vary with the number of teams and the complexity of the product or process.

Vendor Selection

Selection of vendors is a joint responsibility of the value analysis coordinator and the materials or purchasing manager.

Their challenge is to anticipate the kinds of changes the teams might want to consider, so that vendors and experts can be invited who will contribute to evaluating those changes. Since the value analysis system is a creative method, such anticipation is difficult.

Current vendors should account for no more than 50 percent of the invited vendors. The remaining half or more should be catalysts, or so-called hungry outsiders. Some suggested categories are as follows:

Specialty vendors whose products deviate from the garden variety. At the very least, such vendors are likely to stimulate the team members to look beyond existing methods.

Vendors who have been trying to interest the organization in their different approach, but have been unable to address an influential audience.

Any vendor who may have hesitated to present a nonconforming quotation in the past. It is not uncommon for vendors to hold back on ideas that could benefit the product or process because their idea differed from the drawing or specification, thereby excluding those vendors from consideration.

Expert Selection

Selection of experts is the responsibility of the value analysis coordinator. He or she will commonly, however, delegate the responsibility for contacting each expert to a member of top management.

The function the experts perform is identical to that performed by the vendors: 1) to exemplify the great power of reaching beyond the boundaries of the organization, 2) to act as sounding boards for the champions, and 3) to suggest further possibilities for the teams to consider.

Some categories of experts who are worth considering are as follows:

Retired executives or specialists from the organization, or from another.

Top current executives or specialists from another local organization.

An expert from a local university. (Expect to pay a modest honorarium.)

Specialists in a specific functional area inside the organization.

Interaction Session

Vendors and experts are introduced before lunch, then interact freely with the team members until 4:00 P.M. When the workshop includes more than one team, a flip chart is prepared with the names of all vendors and experts and a column for each team. The teams then will determine which vendors and experts they wish to interact with, and the team captains then will check off the appropriate columns (Fig. 9–4).

Six rules guide this session.

1. Only one vendor or expert will interact with a team at a time.

2. The team will remain intact during the interaction. There will be no subcommittees.

3. Each team member will present his or her championed ideas and concepts to the vendors or experts for review and comment.

4. Team members will keep detailed notes on the discussion, including cost estimates.

5. Vendors or experts and team members will exchange business cards to ensure continuation of the investigation.

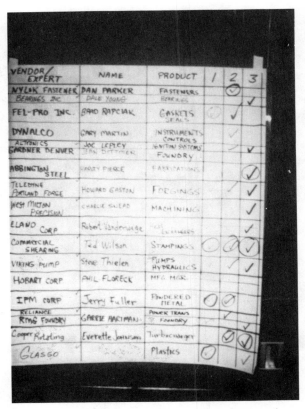

Figure 9–4. A flip chart is used in multiteam sessions to match up each team with the vendors and experts with whom they wish to interact.

6. Each vendor or expert is encouraged to suggest other solutions after the team members have finished presenting their championed ideas and concepts for comment.

With a typical 12 or so vendors and experts per product and with the limitations of one per team at a time and no subcommittees, it would appear that about 11 of them will waste several hours while waiting for their own 20-minute meeting. This is not the case. Each vendor and expert is requested to pull up a chair in an outer ring around a team table and listen to the discussion (Fig. 9–5). Each will thereby become familiar with the team's project and will be far better prepared to interact. Most vendors and experts questioned report that little or none of their time was wasted.

The value analysis coordinator has the responsibility of monitoring the interaction, to prevent any vendor or expert from monopolizing a team or vice versa.

AFTER THE INTERACTION SESSION

At 4:00 P.M., the vendors and experts are thanked and excused. Team members then quickly review their individual plans for investigation of the ideas or concepts that they have agreed to champion. Schedules may be established for brief intersession team coordination meetings.

Figure 9–5. Vendors and experts form an outer circle around the value analysis team while awaiting their turn to interact.

Update Implementation Plan

When a team has completed the elevation of all possible entries in its creative flip-chart lists and have eliminated all nontolerated faults, and when champions have accepted ownership of each resulting idea or concept, each team member then refers to the implementation planning worksheet that was completed on the first day of the study.

Each idea or concept is reviewed against the list of actions required, and the worksheet is updated in view of the characteristics of each active project.

Opening a Dialogue

At this point, a split in viewpoint has invariably developed. The value analysis coordinator, having experienced past successes, is optimistic; but one or more of the team members, while still playing the game, see the entire value analysis system as mostly smoke with little fire.

Even the most experienced value analyst becomes a victim of this difference in perception. It is important to the success of the study that the leader open a frank and forthright dialogue so that the remaining 50 percent of the study may benefit from a common viewpoint. The method for opening this dialogue is to deliver a jolt to the team in the form of the following statement written on the board: Why couldn't we gather together a few great minds and just write anything down?

The team is asked how many agree that such an approach would work as well as the five days of intense, function-based value analysis that they have just endured. The reactions are surprising. It is common for one or more hand to be raised. Invariably, a dialogue is opened.

Take advantage of this openness through an uninhibited discussion, focusing on the insights that the team has successfully created through its function analysis. The major allies of the value analysis process are those who did not raise their hands. A brief "testimony" by a believer tends to weld a team into an impressively focused value analysis machine.

Intersession Effort

Team members are required to dedicate one-quarter of their time between the synthesis and development phases to the investigation and documentation of their championed ideas and concepts.

Each champion will establish a data package on each of the items on his or her synthesis phase worksheet, including all background data, cost estimates, test plans, and so on.

Team members will use their copies of the implementation planning worksheet as reference documents during this intersession period. As the ideas and concepts develop, each champion will update his or her worksheet to reflect changes in the scope and direction of the investigations.

Chapter 10

THE DEVELOPMENT PHASE

OBJECTIVE AND PROCEDURE

In the development phase, the team decides which of the synthesis phase concepts that survived the intersession investigation will be pursued. Those concepts that do survive are entered on final proposal forms, and a complete and detailed implementation data package is prepared.

The development phase is a several-week period during which each champion completes a data package on all proposals that he or she has volunteered to champion. The development session is the kickoff for this phase.

The team captain, using the master list of the synthesis phase worksheet as a reference, asks each champion to give an update on his or her projects. If the champion reports that a project is impractical, it is dropped unless another team member elects to take it on. If the champion reports that a project still has promise, a discussion takes place that leads to recording a brief action plan for the implementation process.

If the project survives this team review, the champion retains ownership until the project is implemented, rejected, or transferred to operations or design engineering for action.

The key element in this data package is the value analysis proposal form (Fig. 10–1). It should be filled out rigorously and in detail. Supporting data such as quotations, savings calculations, labor estimates, and so on should be entered on the reverse of the form, stapled to the reverse, or enclosed in a labeled value analysis proposal folder.

This is the time to pull out the implementation planning worksheet, which was prepared as the first step back in the information phase and updated regularly throughout the study. Use it as a reference document during the preparation of the proposal.

The effort of filling out the value analysis proposal forms is a continuing process. It starts during the development phase session and continues until the final presentation.

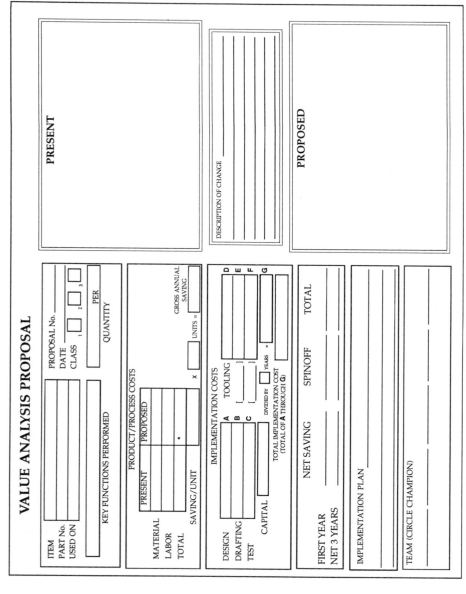

Figure 10–1. *The blank value analysis proposal form is the key element in the data package and is worked on right through the final presentation.*

THE PROPOSAL FORM

The value analysis proposal format is the blueprint for action. Fill it in carefully and completely. It is the key to the accomplishment of the crucial final two steps: approval and implementation.

Include, of course, all data that the champion developed in the initial investigation of the concept or solution, but also include either on the form or in the folder at least the following items:

I. Identification
   A. Name of product or process
   B. Identification number
   C. Quantities
   D. Key functions performed
   E. Class of proposal
II. Costs
   A. Present and Proposed
      1. Material
      2. Labor
      3. Product maintenance costs
      4. Refurbishing costs
      5. Service costs
      6. Tooling and other expenses
      7. Capital cost
         a) Total cost
         b) Predicted savings—1 year and 3 years
         c) Spin-off cost saving
         d) Net present value (NPV) savings
         e) Return on investment (ROI)
         f) Return on assets employed (ROAE)
   B. Narrative
   C. Sketches
III. Implementation plan
   A. Narrative
   B. PERT or Gantt chart
   C. Names of persons required to take action to implement
IV. Backup
   A. Memos, letters
   B. Catalog data
   C. Test results
   D. Vendor quotations
   E. Cost estimates
V. An integral audit format
   A. NPV savings
   B. ROI
   C. ROAE
   D. Scheduled audit dates
   E. Actual savings summary for 1 year and 3 years

Origin of Concept

In the typical value analysis study, a significant portion of the projects that reach the proposal stage represent ideas or concepts that predated the study. All such proposals should be prepared and published—except those that would have been implemented in a timely fashion even if there had been no value analysis study.

A value analysis study tends to sweep up all of the possible product improvements that have idled about in the organization but have not been implemented, for a great variety of reasons. Invariably, although many of the proposals are based on new ideas or concepts that arose during the team study, a substantial portion already existed in some form within the minds or desk drawers of the team members or others.

It is important that these preexistent concepts be given maximum opportunity for implementation. It is equally important that the source of these preexistent concepts be publicly recognized. This is accomplished through a self-adhesive sticker (Fig. 10–2) that is prominently attached to the "Proposed' image block on the value analysis proposal form.

The sticker serves to disarm critics when the results of the value analysis study are made public. When a preexistent idea is simply published on a value analysis proposal form, it is common for people familiar with the idea's history to deride the value analysis effort as having no unique results, but merely stealing the ideas of others and renaming them as value analysis projects.

The Concept of Spin-off

Almost all data entered on the value analysis proposal form is focused on the specific model of the product under study. The single exception is the information entered in the column headed "spin-off' in the cost-saving summary block.

The narrow focus of nearly all the data ensures the validity and auditability of the

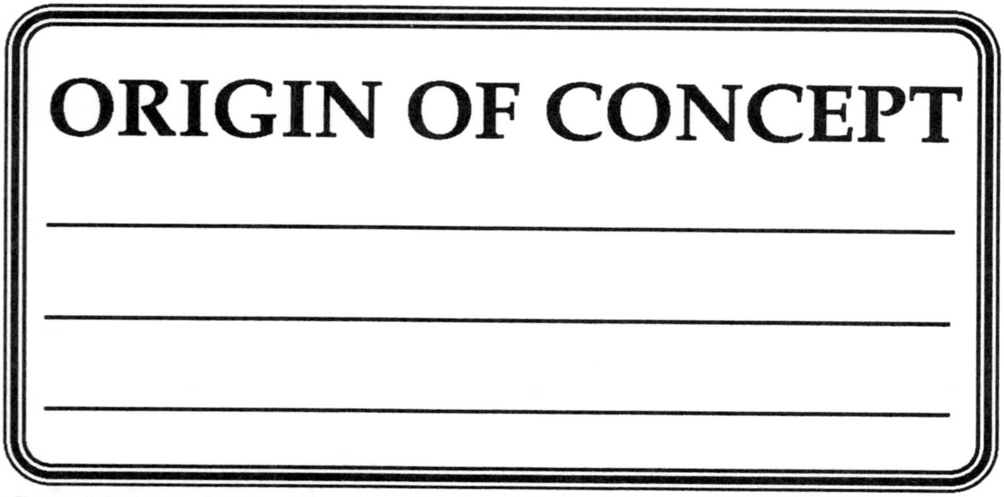

ORIGIN OF CONCEPT

Figure 10–2. The Origin-of-Concept sticker is attached to the "Proposed" section of the proposal form.

results. It eliminates the moving targets and the generalities that characterize so much of the ordinary type of cost-reduction effort—and even some value analysis systems.

Any net dollar saving that results from applying the value analysis change to other models or other products is entered on the value analysis proposal form in the "Spin-off" column. Backup data detailing the specific models and quantities is entered on the reverse of the proposal. Include there a narration and details of any implementation effort required, any effect upon user acceptance, and the calculations that established the net dollar saving.

Class of Proposal

Each value analysis proposal is classified by the champion and the team as a whole. The class of a proposal is related to the ease of implementation and thus relates directly to the length of time before the proposal can be implemented.

The three classifications are described below:

Class 1 proposals: clean, ready to go, easy to implement. No testing. Marketing approval only a formality. No significant drawing changes required. Proposals are often implemented during the study. Payback: significantly less than 12 months.

Class 2 proposals: look good to team, but need testing and/or marketing approval. Payback: significantly less than 24 months.

Class 3 proposals: major departure in design; very large capital investment. Very long lead times. Company policy changes. Payback: longer, but within hurdle rates.

Backup Data

On the reverse side of the value analysis proposal form is an area marked "Backup data." Here should be recorded any details of the calculation of "Spin-off" savings. Include data here on the before-and-after costs, implementation costs, identity of applicable models, and notes on a plan for implementation of the "Spin-off" redesign.

This area also serves to record the names and locations of vendors or experts who have contributed data to the investigation, as well as quotations from outside vendors and in-house labor and material cost estimates.

A third area to be recorded here is the detail of any engineering or economic calculations to support the proposal. The reverse side of a value analysis proposal form is shown in Figure 10–3.

Any additional documentation should be stapled to the reverse of the proposal form or kept in a labeled folder.

Review and Audit Control

The reverse of the form also carries the structure for a system of follow-up to ensure implementation (see chapter 12). The champion leaves the three "Review" entries blank. The value analysis coordinator will enter the schedule review dates when the total proposal package is prepared for final presentation to management. The section marked "Audited results" will be filled in by the comptroller's office when the proposal is implemented.

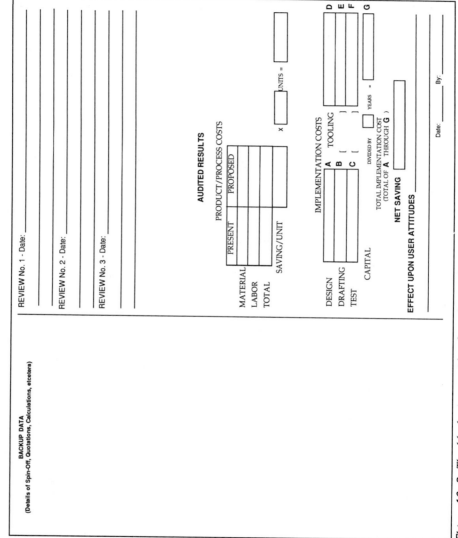

Figure 10–3. *The blank reverse side of a value analysis proposal form is used for backup data, including information about vendors and experts, and a variety of detailed cost estimates and calculations.*

THE GANTT CHART

Each value analysis proposal must be accompanied by a thorough schedule of the activities that must be performed to ensure implementation. The ideal format is the Gantt chart (Fig. 10–4).

The champion is responsible for filling out the list of activities on the Gantt chart, with the guidance of appropriate authorities. For engineering and drafting activities, the authority is usually the project engineer, while tool design data will come from methods engineering. Cost and timing of a market research study will be developed by marketing.

Each of these authorities will be initially reluctant to record the activities, manloading, and elapsed times without detailed data on the redesign project and a considerable period for study and calculation. It is essential, however, that a Gantt chart be prepared at this time, even though a rigorous basis for estimation does not exist. It is therefore the responsibility of the champion to enter the best estimate of activities, times, and manloadings, based upon his or her discussions with the appropriate authorities.

After an initial Gantt chart has been drafted, the champion will review it with each authority to ensure that it is approximately representative of the required implementation process.

UPDATE PLAN FOR IMPLEMENTATION

When the team has completed its review of all championed projects from the synthesis phase, each team member then refers to the implementation planning worksheet, which was completed on the first day of the study and updated at the completion of the synthesis phase session.

Each project is reviewed against the list of "actions required," and the worksheet is updated in view of the characteristics of each active project.

THE MELDING OF RESULTS

All of the synthesis and development phase effort to this point has focused on the individual champions and their independent proposals. This is intentional and appropriate.

In taking full advantage of the awesome power of individual minds, the value analysis system results, quite naturally, in a group of individual proposals. The team sessions tend to dilute this individuality by forcing a continuing discussion, which ensures that all of the proposals are at least compatible. Under certain conditions, however, a complete integration of proposals is essential. Such situations include the following:

An upstream value analysis study being conducted in the concept stage of product design

A complete product redesign that will include the value analysis study results

A group of value analysis change proposals that are highly interrelated and therefore require melding and integration

PROJECT	*Head Casting*		PROJECT No.	*3-2-5*
PREPARED BY	*S. Grass*		DATE	*12/10/88*
LEGEND:	EVENT: ∧	ESTIMATED	ACTUAL	RESCHEDULED
[Enter the labor hours and type of labor to the right of each activity]		START > FINISH <	START > FINISH <	START ⊙> FINISH ⊙<

1986

	MONTH		March		April		May		June		July	
	WEEK											
A	*Draw Machined Head*	> <			20 hour Design, 20 hour Detail							
B	*Draw Casting*	> <			10 hour Design, 15 hour Detail							
C	*Draw Plate & Assembly*	> <			3 hour Detail							
D	*Quote Casting*	>			<		6 hour Purchasing					
E	*Estimate Machining, Assy*	><			12 hour Manufacturing Engineering							
F	*Issue Purchase Order*				∧							
G	*Receive ten Castings*											
H	*Machine five Castings*											
I	*Test*											
J	*Release to Manufacturing*											
K												
L												
M												
N												

MO	*August*		*September*		*October*			
WK								
A								
B								
C								
D								
E								
F								
G		∧						
H		>		<	86 hour Model Shop			
I			>		<	40 hour Test		
J				∧				
K								
L								
M								
N								

Figure 10–4. Ten activities related to the heat pump proposal are scheduled and manloading noted on this Gantt chart.

In such situations, the team schedules a formal, full-day team melding session, approximately one week after the development session and before the final presentation to management. The objective is to integrate the results of the development phase into a single optimized system, with maximum fulfillment of the users' needs and wants at minimum cost.

The focal point of this melding session is commonly a wall-size diagram, prepared in advance, containing a blank block for each required element of the design and structured with a logical flow. The individual champions are each challenged to flesh out and/or modify this block diagram by entering into the appropriate blocks a concise description of each of the proposals that the champion feels is appropriate. An interactive team discussion gradually synthesizes a complete and optimized system (Fig. 10–5). The melding session is led by the suitably briefed project engineer on the team.

The completed block diagram contains notations on cost and user need fulfillment, as well as explanatory notes on each block and interface. Where the team feels that several alternative systems deserve investigation, each is noted on the final block diagram.

A summation of the collective conclusions of the team, the diagram should be redrawn after the melding session, using a similar format and including all notations and sketches, but in a size that will permit each member to use it as a constant reference during preparations for the final presentation to management.

Figure 10–5. Team members work on a huge block diagram on the wall in order to put together a complete and optimized system.

DECISION MAKING

The challenge of the development phase is to analyze all proposals and then select the best solutions. At this stage in the value analysis system, we are commonly faced with a new kind of problem: We have several optional solutions to the same problem. Which one is best?

We are faced with hundreds of similar situations each day of our lives: What shall I wear today? What shall I have for breakfast? Shall I buy gas on the way to work or later? What shall I work on first? Next?

None of these questions is earthshaking in its implications, but collectively the decisions we make each day significantly affect the outcome of the day for us, and often for those with whom we interact.

Indeed, the collective quality of those decisions, each day, each week, each year, defines our total effectiveness as people.

If we can assume that each of us is, indeed, effective in our life and in our career, then we must have a pretty good system for making those hundreds and thousands of decisions.

Just what is that system?

What logical process do we activate when we are forced to make a choice between two or more alternative actions?

"I look at all the options and I choose one. It usually turns out to be the right one."

"I consider all the variables carefully, and the answer is usually obvious."

"I just do what feels right to me."

These are actual responses from people who are successful managers. These people make good decisions, but they don't know what process they are using.

A Science of Decision Making?

In the mid-1950s, the new management science of operations research and the arrival of the high-speed computer triggered at least a dozen automatic decision-making systems. Each was touted as both easy to use and rigorous. Most were neither.

In the early 1960s, the first truly easy-to-use and rigorous system was developed by the late Carlos Fallon, manager of value analysis at RCA (Fig. 10–6). Fallon was a rare combination of down-to-earth practitioner, mathematician, philosopher, and mesmerizing public speaker. His insight gave us Combinex©, which he published nationally in his book *Value Analysis to Improve Productivity* (Fallon 1971). Fallon's work has been modified in only minor respects in the description below and on the following pages.

COMBINEX©—A SYSTEM FOR METHODICAL DECISIONS

by Carlos Fallon
(Modified for direct magnitude by T. C. Fowler, CVS)

By using the Combinex© system, you can objectively determine which of your solutions

Figure 10-6. Carlos Fallon is the creator of the Combinex® process for decision making.

are truly the best from an objective viewpoint, and prove it to those who will make the ultimate decision to implement.

While Combinex© is an easy-to-understand, easy-to-administer, orderly process for making decisions, its real power lies in its ability to show clearly how the answer was obtained, and how the attitude of the decision makers helped to shape those results.

The following five steps define the Combinex© process. Each is explained in detail on the following pages.

1. Select criteria.

2. Define limits, reference level, and sensitivity curves.

3. Establish weights, or relative importance factors.

4. Rate each alternative on each criterion.

5. Calculate relative "figures of merit."

The forms shown in Figures 10–7 and 10–8 will guide the analyst through the procedure. Excerpts from these forms illustrate the process as described below.

Step 1: Select Criteria

To start the evaluation process, first select the criteria that will affect the outcome.

Evaluative criteria are simply a set of objective standards by which to judge the worth of alternative solutions to a problem. Three considerations control the selection of criteria.

First, criteria must ultimately be selected by the decision maker, that is, the person who has the authority to approve the final selected alternative solution.

Second, criteria must be measurable. Ideally, they should be parameters. Some examples of measurable parameters are as follows:

Performance (in terms of measurables such as reliability in MTBF, maintainability in MTTR, level of accomplishment of basic or supporting functions)

Convenience, attractiveness (specific elements, measurable through market research)

Cost (in terms of level of investment required, manufacturing cost, service cost, total cost, ROI, net present value, cost versus target, cash payback, and so on)

Risk (where it can be quantitatively appraised)

Effect on safety

Feasibility

Third, criteria must be independent, that is, mutually exclusive. Use the following means to identify key criteria:

Freewheel. Temporarily defer judgment as in a creative problem-solving session. Record all conceivable criteria, as shown in Figure 10–9. If you are not the decision maker, attempt to second-guess the person who is by putting yourself in his or her place.

Judge. Through combination and evaluation, select a group of criteria that are sufficiently broad, measurable, and mutually exclusive to represent a best estimate of the measures by which the decision maker will evaluate the alternatives. Remember that in the end, the decision maker will have to agree with your choices.

Record. Record the name of each selected criterion on a form similar to that shown in Figure 10–8, using as few words as possible.

Define. Define the selected criteria by recording, on each form, those constraints necessary to make each criterion clearly mutually exclusive. Add any notes to ensure that each person involved in the decision and each person who may evaluate the results will perceive each criterion in the same frame of reference.

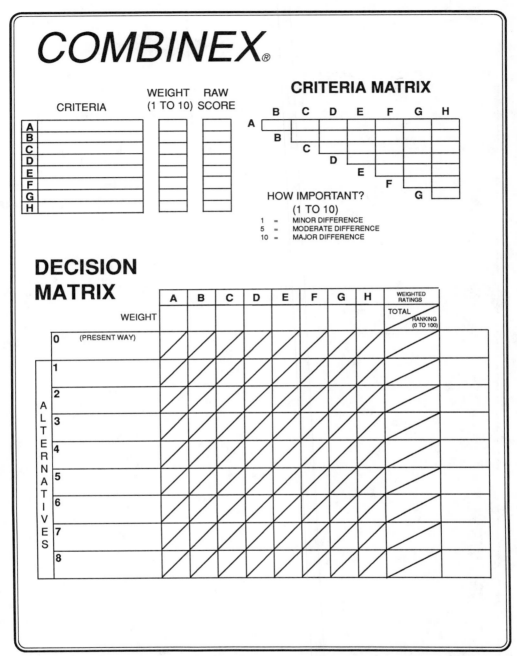

Figure 10–7. The Combinex® decision matrix form is an important part of this decision-making process.

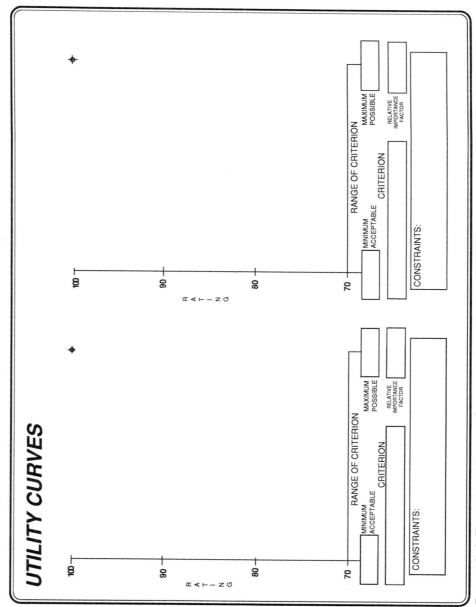

Figure 10–8. The criteria utility curve form is valuable in Combinex® evaluation.

Figure 10–9. The team leader stands next to a board filled with suggested evaluative criteria for a Combinex® session.

Step 2: Define Limits, Reference Level, Sensitivity Curve

This step comprises five tasks. First, set upper and lower limits to establish the range over which valid decisions can be made. The lower limit is the minimum acceptable, the lowest level or worst condition that can be permitted. The upper limit is the maximum possible, the highest level or best condition that can effectively be used, or the most available, whichever is less.

Occasionally, the minimum acceptable is numerically high, while the maximum possible is numerically low. An example is where product cost is the criterion. Here, the highest level or best condition for the criterion is clearly the lowest dollar cost.

Second, establish a "reference level." This aids in defining the curve and simplifies the tasks of weighting the criteria and rating the alternatives.

The reference level can be one of the following points:

- midpoint—center of the range

- typical—usual value of the criterion

- present value—current or known value of the criterion

Use whichever point is most significant.

Occasionally, there is no present value, and neither the term *midpoint* nor the term *typical* has any valid meaning. In such a case, the reference level is ignored and the limit values described below are used as references in performing the weighting of the criteria and the rating of the alternatives. A detailed example of the utility curve of the criterion "efficiency" is shown in Figure 10–10.

Third, enter the limits and reference level. Establish a horizontal and vertical scale on the criterion form such as the one illustrated in Figure 10–10.

Enter on the abscissa (x-axis) the lower and upper levels of the value of the criterion. Also enter on the abscissa the reference level of the value of the criterion, if one has been set. The range of the ordinate (y-axis) is 70 to 100 to imply a typical rating range from "just passing" to "outstanding."

Fourth, define the curve. Add to Figure 10–10 a curve that describes the variation in the influence of the criterion between its limits. In many cases, a linear relationship exists between the acceptability of an alternative solution and different values of the criterion. In others, the relationship is nonlinear, or even discontinuous. Develop a curve describing this relationship throughout the 70–100 range of the standard scale.

Establish the shape of the curve through a combination of two methods. First, determine the rate of change of the criterion at three points: the lower limit, the reference level, and the upper limit. Simply estimate the effect on the rating of a small change in the value of the criterion at each of these points. This defines three slopes from which a curve can be created. The other method is to rate the criterion at several different values. Draw a "best fit" curve through the points.

Fifth, review criteria. At this point, if you are not the decision maker, review the criteria and curves with that person. Do not proceed until the decision maker has agreed that the data represents his or her criteria and curves. There are two important reasons for this: It is not possible to place oneself completely in another person's place, and the

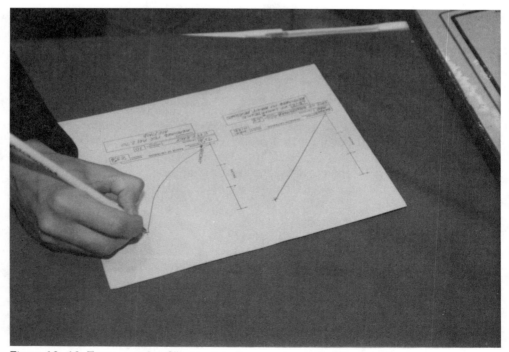

Figure 10–10. Team member filling in data on a preprepared utility curve form.

decision maker later will be more inclined to accept your recommendations if he or she is involved at this point.

A group of decision makers defined on a flip chart a list of 8 criteria, reduced from 19 originally through a process of combination and redefinition based on the fundamental principle of mutual exclusivity. These are the criteria that next will be weighted:

- investment required

- risk

- manufacturing cost

- minimum size, weight

- efficiency

- field adaptability

- ease of installation

- ease of maintenance

Step 3: Assign Weights

In the Combinex© process, two steps virtually determine the final outcome: the selection of the criteria and the relative weighting of those criteria. Of the two, criteria selection presents the fewest problems.

When the relative importance or weighting of criteria is established on the basis of simply the best judgment of the person or team performing the evaluation, three sources of error are always present.

One is provincialism, or self-interest. This is the error caused by the different frames of reference or viewpoints of the evaluator and the decision maker. Essentially, this is partiality, bias, prejudice in favor of whatever is familiar to each. Provincialism is obviously not present when the decision maker is personally establishing the relative weighting.

Another is subjectivity, or emotion-based preferences. Their effect can be minimized with effort, but never eliminated.

The third is complexity. When more than two criteria are being compared, the process requires the simultaneous mental rating of a very large number of factors. The human mind is not capable of rationally interrelating so many permutations.

A technique that eliminates the complexity and greatly minimizes provincialism and subjectivity is called paired comparisons.

This method initially assumes that all criteria have equal importance. Each is then forced in turn into a comparison with each of the others. With each pair thus formed, a judgment is made as to which is more important to the decison maker. An estimate is also made of the magnitude of the difference in importance.

These comparisons are made within two frames of reference, and the final decision is made on the combined bases of both comparisons.

In one method, compare each pair at the reference level, if one has been set. Hypothesize an equal decrease in the value of each criterion from its respective reference level. The more important criterion is the one in which this decrease most affects the acceptability of typical alternative solutions. Repeat the procedure, hypothesizing an increase in the value of each.

In the other method, compare each pair at the upper and lower limits. Perform the above evaluation at the upper limit, hypothesizing a decrease. Repeat at the lower limit, hypothesizing an increase.

Reference-level comparison (the first method) is the better one to use, particularly when the reference level represents the typical or present value.

The results of the forced-pair comparisons (A to B, A to C, and so on) are calculated and entered to give a relative importance factor, or weight, for each criterion (Fig. 10–11).

The procedure for this is to enter the letter of the criterion judged to be most important at each intersection on the weighting matrix. Add to that letter a number (1 to 10) signifying how much more important. If both are equally important, enter a zero.

For example, the first matrix entry in Figure 10–11, "A-8," is at the intersection of row A and column B. This indicates that criterion A is more important than criterion B by a factor of 8 (on a scale of 1 to 10).

The raw score shown in the rectangle on the right is the sum of all the As (35), Bs (27), and so forth.

If any raw score is zero, enter, as the raw score for that criterion, an amount equal to one-half of the lowest value shown for any of the other criteria.

The weight is determined by rescaling the raw scores into a 0 to 10 scale. The criterion with the highest raw score is given a weight of 10, while the others are proportioned to

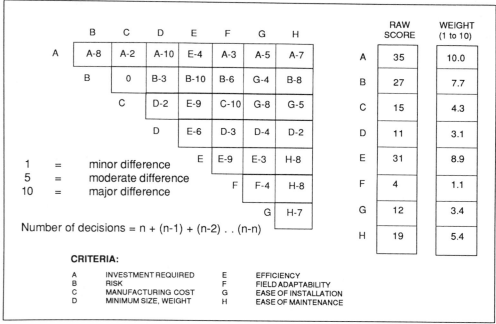

Figure 10–11. This typical paired comparisons criteria weighting form is filled in to represent a typical set of data involving eight previously determined criteria.

that highest score. For example, the weight of criterion C, manufacturing cost, in Figure 10–11 is calculated as shown in Figure 10–12.

Step 4: Rate Each Alternative on Each Criterion

The team will then use a decision matrix (Fig. 10–13) to compare and evaluate alternative solutions.

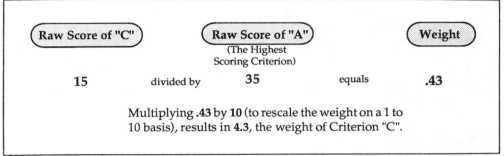

Figure 10–12. In this example, the formula is used to calculate the weight of criterion C, manufacturing cost.

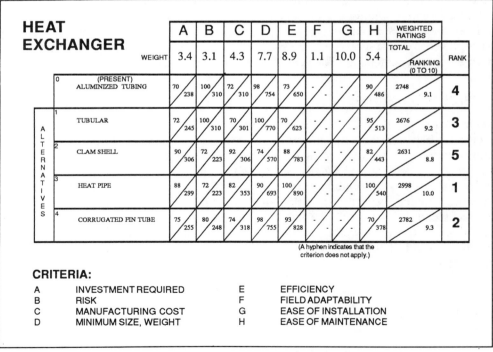

Figure 10–13. A completed decision matrix rates four different approaches to a heat exchanger problem.

To create this matrix, list the criteria with their associated weights across the top and list alternative solutions down the left.

Next, rate each alternative against each criterion, entering the rating (70 to 100) in the appropriate block. Ratings are obtained by picking values from the sensitivity curve.

Step 5: Calculate the Relative Figures of Merit

Multiply each rating by the weight for the criterion, and enter the result in the same block. Separate the unweighted and weighted ratings with a diagonal line.

Total the weighted ratings horizontally for each alternative. The result is the figure of merit for that alternative.

Finally compare and select alternatives with high figures of merit. Note that a very small difference in the numerical figures of merit (on the order of 2 percent) is not significant, but as the difference approaches 5 percent it becomes significant. Avoid the tendency to choose an alternative with a significantly lower figure of merit simply because it instinctively feels better.

Closing Comments

While the alternative with the significantly higher figure of merit is, initially, the so-called winner, the true power of the method stems from the unparalleled visibility of the data. On a single sheet are displayed all three of the factors that contribute to the final choice: criteria, weighting, and rating. Analysis of the anatomy of the data often develops an unprecedented insight for both the value analysis team and the decision maker.

In addition, Combinex© is a powerful selling tool. It closes the loop with the decision maker. Recommendations to a top manager that are presented in terms of his or her own acceptance criteria are far more likely to be well received.

INTERSESSION EFFORT

Team members are obliged, as in the last intersession period, to apply one-quarter of their time between the synthesis and the presentation phase sessions in investigating and completing their data packages for the final presentation to management. The key document for collecting and integrating the data is the value analysis proposal form.

The value analysis coordinator is responsible for planning the final presentation and preparing the final report of the value analysis workshop. He or she will set a deadline for completion of the value analysis proposal forms and will make reduced copies of the proposal forms as well as prepare summaries and other material for inclusion in the final report. The coordinator also will schedule any required intersession team planning meetings.

Chapter 11

THE PRESENTATION PHASE

The value analysis workshop team has developed, verified, and documented a number of proposals for change to the product or process under study. The effort involved in reaching this point has been both creative and intense.

It is now time to implement these changes, an effort that requires even greater creativity and intensity.

The presentation of results to management is the climax of the value analysis workshop, but it does not by any means signify the end of the effort for team members. Each proposal has a champion who has accepted the responsibility for carrying his or her solutions to the point where they can be dropped, implemented, or turned over to an appropriate person for further investigation and eventual implementation.

The final presentation session is, in fact, the first day of a sometimes lengthy implementation process.

SELLING THE PROPOSALS

The objective of the final presentation is to sell to operating management an integrated package of change proposals. The hoped-for "buyers"—the audience at this presentation—include the chairman and members of the value analysis council; in addition, any member of management who has any involvement with the product under study should be invited.

It has been said that if a proposal is a really good one, it doesn't require selling; it will be implemented simply because it is a technically superior solution. This is just not true. Even the most logically obvious proposal for change requires some sort of approval. The dynamic involved in that approval, while partly based on the technical merit, is primarily an emotional and interpersonal matter. It is appropriate to apply just as much ingenuity and diligence to the selling of a proposal as was originally applied to its development.

PREPARING FOR THE PRESENTATION

Rehearsal and Verification

During both of the three-to-four-week intersession periods of the value analysis job plan, it is common for the team to meet together on a fairly regular basis; once per week is a recommended minimum.

In addition, it is necessary for each team to schedule a dress rehearsal several days before the final presentation to management, which, after all, is just a two-to-three-hour focused summary of results. Since it is essentially a selling session, it must be condensed, pointed, and brief. Rehearsal serves to remove the rough edges, eliminate the nonfunctional verbiage, and integrate the champions' individual presentations into a unified whole.

The Implementation Planning Worksheet

In their first session, the team members recorded on their implementation planning worksheet a series of possible areas of change. For each area, a list was recorded of actions required to ensure implementation of change.

The team updated this worksheet at the completion of the synthesis phase session, and further refined and updated it at the completion of the development phase session.

In planning the presentation to management, this completed implementation planning worksheet (Fig. 11–1) is a key reference document. The insight that results from a critical review of this document often suggests the most effective method of attack for structuring the presentation and selling the proposals.

Program Evaluation Form

Immediately before each team presents its results to management, its members should each fill out a program evaluation form, a questionnaire similar to that shown in Appendix B (see Fig. B–3). This is an essential step in the constant effort to improve the method and content of the value analysis study process. The data gathered at this critical point in the process is often uninhibited. It is also far more useful in optimizing the process than any introspective process of post-program soul-searching by the workshop staff.

THE RIGHT APPROACH

Each champion must actively participate in the presentation of each of his or her Proposals (Fig. 11–2). Success in "selling" them depends on the right approach as much as on the quality of the proposals themselves.

Don't argue. Stay self-assured. Remember, you have become the expert on the problem and on the solution you are proposing.

Minimize the threat of change. Change involves risk, the possibility of embarrassment. Respect the chain of command. Don't infringe on anyone's status, real or imagined. "Business decisions, in the main," said Vic Short, "are based on fear of personal loss."

Anticipate both sides of the story. Neutralize opposition by recognizing other ap-

Possible Areas of Change	Actions Required to Ensure Implementation of Change
1 *Manufacturing Process*	*Verify with manufacturing engineering - Jim Jackson, who designed the manufacturing process and many of the specialized machines.*
2 *Refrigeration Process*	*Ed Nelson designed the system used in the 4700. He has indicated that he is open to suggestion, but we should check in with him early on any significant modification.*
3 *Sourcing of parts*	*Production is allocated to both the satellite plants and certain prime vendors. Any proposal to pull work to the main plant should include a plan to balance out production.*
4 *Market Considerations*	*The marketing department feels that the model 4700 is their flagship and should remain inviolate. Have periodic sessions with the Product Manager.*
5 *Reliability and ruggedness*	*Haskins of Reliability Engineering has a passion for large safety factors. Back up all proposals that involve reducing gauges of material with unchallengeable data.*
6 *Motors and electronics*	*The Ithaca R&D Center has design cognizance on all electronics and associated control circuitry. Give them regular updates in person or by phone as work progresses.*

Figure 11–1. An implementation planning worksheet for the "Theme Thread" heat pump is filled out listing six possible areas of change and the actions required to implement them successfully.

proaches. Admit and quantify the possibility of failure, but emphasize the risk of *not* making the change.

Disarm critics. If the proposal is not original, clearly acknowledge its source. Do not claim that the value analysis workshop was responsible for creating the concept if it was, in fact, a preconceived solution from another source.

Legitimize the proposal by citing the experience of others in using the same or a similar concept.

Head off objections. Be prepared to respond to at least the following roadblocks:

Figure 11—2. The project engineer, champion of one of his team's proposals, displaying a working model of it to the president and his staff during the final presentation.

- It isn't in the budget
- It won't work here
- It's been tried before
- We don't have time
- The price is too high
- It's not practical
- It would make our products obsolete
- It's ahead of its time
- It's not our problem
- Why change? It works!
- The boss would never go for it
- It's against policy

Call for action and make it easy for others to act.

Promise Only the Probable

A technique that evolved as part of the original format of the value analysis training seminar has often resulted in negative management reaction. At the completion of the team presentations, the value analysis coordinator would announce the total proposed savings. The figure quoted was often dazzling to top management and served as an impressive factor in the selling of the value analysis system.

The negative reaction usually took place several months after completion of the value analysis study. In a typical scenario, executive management is faced with a weak quarter and calls in the value analysis coordinator to determine what portion of the $8 million of proposed savings will return quickly to the bottom line. Executive management is typically told that only a small percentage of the proposals have been implemented, but that more is expected. When the value analysis coordinator explains, for instance, that "$150,000 has been implemented," he or she is really telling executive management that the value analysis system has "lost" $7,850,000 that was promised several months before. Credibility suffers, program support fades, and the prospect of value analysis becoming a way of life in the organization is set back.

Modern value analysis insists that the only dollar savings reported at the final presentation be *probable* savings. All savings and all implementation costs are estimated realistically but conservatively, and the only net dollar savings reported are those that are already implemented or are highly probable of being implemented.

PROCEDURE: THE TECHNIQUE OF EFFECTIVE PRESENTATION

Thousands of books, articles, and "magic formulas" have been written on the psychology and dynamics of selling your proposal. None of them captures the essentials better than the system the General Electric Company chose: the four-point formula that Richard C. Borden called Effective Presentation.

Following is an excerpt from Borden's book, *Public Speaking as Listeners Like It* (Borden 1935). The examples Borden used in 1935 have been updated and rephrased in value analysis terms. The system of Effective Presentation has far broader application, however, than in the presentation of value analysis study results; the reader will find it valuable in any formal presentation. It even provides an effective structure for selling a point in an informal discussion.

Step 1: "Ho-hum!"

In the first section of your presentation, start a fire! Kindle a quick flame of spontaneous interest in your first sentence.

Smokers do not like matches that fail to light with the first scratch. Listeners do not like speakers who fail to "light" with the first sentence.

When you step to the front of the room, do not picture your audience as waiting with eager eyes and bated breath to catch your message.

Picture it, instead, as definitely bored—and definitely suspicious that you are going to make the situation worse.

Picture your listeners as looking uneasily at their watches, stifling yawns, and giving vent to a unanimous "ho-hum!"

The first sentence of your presentation, like the Indians who bite the dust in the opening scene of an old Western movie, must crash through your audience's initial apathy.

Don't open your presentation of the results of your value analysis study by saying, "Our team worked on the P-14 smoke alarm." Say instead, "Last month in St. Louis, 40 houses burned to the ground."

Preacher Henry Ward Beecher lighted up the audience one night. He rose to face a hall full of dozing churchpeople and began, "It's a God-damned hot day."

A thousand pairs of eyes goggled and an electric shock brought the audience erect. Beecher paused, raised a finger in solemn reproof, and went on, "That's what I heard a man say here this afternoon!"

He then launched into a stirring condemnation of blasphemy—and took the crowd with him.

So: Start off with a ho-hum crasher.

Step 2: "Why Bring That Up?"

In the second section of your presentation, build a bridge! Your listener lives on an island—an island of his or her own interests.

Why should anyone be concerned, for instance, about the burning houses in St. Louis that you have introduced as the subject of your presentation?

"Yes," your listener admits, "you caught my attention with your intriguing opening sentence. But in the cold light of second thought—why bring this subject up anyway?"

The second section of your presentation must squarely answer this question:

"I bring up this subject because, if all 40 of those houses had been protected with our P-14 smoke alarm, we would still have lost two to fire.

"Our P-14 alarm has a 5 percent field failure rate. That means 2 out of 40 fail to perform their advertised function. This lack of dependability has seriously eroded user confidence in our entire line. Shipments are down.

"Each of us is vitally interested in improving our reputation for reliability. The value analysis workshop team has attacked and solved the problem of the P-14 failure rate."

Whatever the subject of your presentation, you must build a bridge to your listeners. Until this bridge is built, you are not ready to begin the body of your presentation.

Step 3: For instance!

In the third section of your presentation, get down to cases!

Let's assume that you have introduced your presentation in an interesting way, arresting all ho-hums with your first sentence, and that you have deftly convinced your listeners in your second sentence that the subject strikes their interest. Now get down to specifics.

The body of your presentation must be keyed to one relentless demand from your audience: "For instance!"

At this point, maintain control. Do not present your points in a jigsaw jumble. Offer the proposals of your team as organized platoons—in marching order:

"Modification of the battery holder and its contacts increases its reliability from 4,200 to 19,400 hours between failures and reduces cost from 86 to 21 cents per unit.

"Changing the trigger circuit to a custom-integrated circuit increases reliability from

374 hours to 60,000 hours and reduces cost of labor and material from $2.20 to 62 cents with a design-and-tooling cost of $6,200. Break-even on this investment is three months.

"Shipping damage accounted for 35 percent of field failures. Addition of a foam shock mount internal to the smoke alarm isolates the delicate circuitry and sensor from shipping and installation damage. This change alone decreases the field failure rate from 5 percent to 3.5 percent and increases the unit cost by only 4 cents per unit."

Note that when one platoon marches by, that's the end of it. The speaker does not

say, in the middle of discussing the foam shock mount, "Oh—by the way, I should have mentioned that the integrated circuit we are suggesting is also more resistant to shock."

Remember this when you come to the body of your presentation: Listeners like speakers who serve their "for instances" as full-course dinners, not as goulash!

Step 4: "So what?"

In the concluding section of your presentation, ask for action!

The end of your presentation, like the end of your pencil, should have a point.

The conclusion of your presentation must be more than a graceful leave-taking. It must be more than a review of the points covered in the body of your presentation. It must be more than a reminder of your subject's general importance.

It must answer the audience's question: "So what?"

"So far so good," say your listeners. "You have introduced your presentation in a

manner to attract our attention. You have built a bridge to command our serious attention. You have illustrated it with convincing proposals. But now what? Where do we go from here? What do you want us to do about all this?"

In the conclusion of your presentation, ask your audience for a specific action—some response that is within their power to give:

- Approve!

- Test!

- Invest!

- Buy the machine!

- Assign the engineers!

- Sign off the redesign proposal!

When you feel tempted to end your presentation without such a call for action, remember the Chinese proverb of the Middle Ages: "To talk much and arrive nowhere is like climbing a tree to catch fish."

VISUAL AIDS

Since the old saying "One picture is worth 1,000 words" still holds; use visual aids to sell your concepts and solutions. There are several effective kinds of visual aids.

Figure 11-3 shows a number of effective ways to focus the attention of the attending managers on this team's $19,600 target. First, the large image of a thermometer was

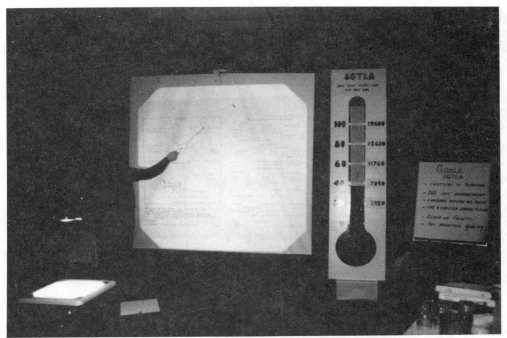

Figure 11—3. A massive thermometer at the front of the room highlights the presentation of proposals; the red band is moved up as champions complete their proposals. The fact that the team exceeded the goal marked on the thermometer is thereby dramatized.

designed so that the solid red band could be moved upward as each championed proposal was presented. Such an element of showmanship is appropriate as well as effective.

In addition, the champion shown is presenting one of his value analysis proposals through the use of an enlargement of an 8-by-10 transparency of the proposal form. He is also using a pointer to emphasize one element of the "before" image.

Further, on the right is a card reminding the audience of the six goals that operating management established at the workshop kickoff.

Add impact using such visual aids as actual photographs, graphs, cartoons, cost-and-savings summaries, and line drawings. Professional artistry is not necessary, but the better the visual aid, the greater the impact.

The easel poster shown in Figure 11—4 was displayed at the completion of the final presentation. It served as a helpful reminder in an open discussion between the team and the management group.

A touch of light-hearted showmanship is often effective. Figure 11—5 shows a team member taking make-believe money out of an attaché case and adding it to a growing pile in front of the top manager. Other appropriate theatrics have included taking a hat, placing it on the table in front of the president, and having each champion toss into it coins equal to the dollar savings from each proposal.

Figure 11—6 illustrates a technique sometimes referred to as "Ta-dah!" At the conclusion of the value analysis team presentation of results to Hobart Brothers Company management, a team member, with a flourish, removed a box covering the redesigned

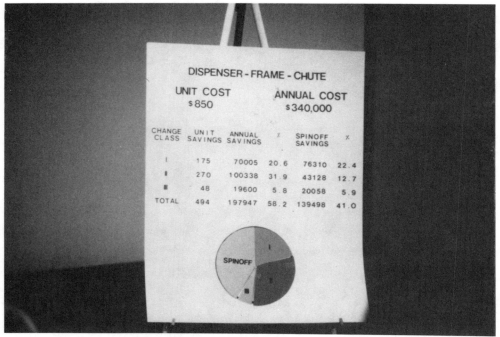

Figure 11–4. An easel at the front of the room supports a poster showing a typeset pie chart with numbers that summarize the team's results, by class of proposal.

battery charger. This dramatic approach could be regarded as show business inappropriate to a serious business environment; in our experience, however, it is universally well accepted and is in keeping with the fundamental concept that even a great idea must be sold.

Models are among the most effective selling tools. For a process change, they may be merely "paper models"—that is, forms or flow charts. For a product, they are often simulated models made from foam board or manila paper. Where possible, of course, an actual working model of the proposed configuration is most effective, but even mock-ups can be very useful.

A model of the new design is often the only way to convey its advantages. In Figure 11–7, the device on the left is the present design of a hand control for an X-ray machine. On the right are two stages in the construction of a mock-up of the redesigned control. The key advantage of the new device is one-handed operation.

Often a team prepares a complete and detailed scale model and displays it after presenting all of the value analysis proposals. A poster showing the key features of the new design can supplement the display of the detailed scale model (Fig. 11–8).

Another important visual aid for each proposal is the Gantt chart, which displays the actions required of each of the organization's support functions, including the time-phasing and man-loading of each activity. The believability of the presentation is greatly enhanced by referring to key actions required for implementation.

The key visual aid, however, is an overhead transparency of the value analysis proposal form. Use it to supplement your selling effort with well-ordered, documented facts.

Figure 11–5. A team member opens an attaché case full of "money" at the end of each champion's presentation, flipping the appropriate portion of the contents onto a growing pile in front of the vice president.

Sometimes it is most effective to reveal the visual message one step at a time in synchronism with the presentation. This can be done by covering some elements of a presentation poster or flip chart with blank strips and removing each at the proper point in the presentation. Another way is to hinge strips to the edges of an overhead projector and either flip each into place or remove each strip at the appropriate point in the presentation to focus the audience's attention where you want it (Fig. 11–9).

Be creative. Demonstrate enthusiasm for your proposals. Apply some of the same energy to their presentation that you devoted to their preparation.

THE PRESENTATION SUMMARY

When the members of each value analysis team have completed presenting their championed proposals, the team captain presents a summary of the proposals. The visual aids for this summary are integrated with the team members' individual proposals; they will vary with the needs and characteristics of the audience and the creativity of the captain and his or her team. Typical visual aids include the following:

A list of the financial data for each proposal

A summary of financial and user acceptance results

A time-phased listing of the required investment

Figure 11–6. A team member stands behind "before" and "after" models of a battery charger, unveiled with a flourish during a final presentation.

For the "Theme Thread" product, the visual aids are Tables 11–1 and 11–2 and the following financial summary:

Savings per unit
    Class 1 $16.22
    Class 2 $40.55
    Class 3 $404.06
    Total $460.83
                              (38% reduction)

Gross annual saving $3,382,901
Investment required $1,105,595
Return on investment, first year   306%

All three were also published as part of the team's written final report.

The financial summary and results summary are self-explanatory. The third visual aid is a predictive summary of the time-phased investment that management is being requested to authorize. The importance of this document can be appreciated when we reflect that management is properly skeptical about promised rates of return of 200 to 300 percent or even more on their investment through value analysis. It is important that the teams present data that is clearly realistic, both in costs to be saved and in investment required.

The investment summary (Table 11–2) can be prepared by hand, with some effort,

TABLE 11–1. VALUE ANALYSIS TEAM RESULTS SUMMARY FOR MODEL 4700 4-TON HEAT PUMP

Proposal		Unit saving	Gross first-year saving			Implementation cost	Net saving	
			Project	Spin-off	Total		1st yr.	3 yr.
	Class one proposals							
2-1-1	Compressor mount	$1.64	$38,792	$9,840	$48,632	$5,400	$43,232	$135,937
2-1-26	Simplify access	($.25)	($4,370)	—	($4,370)	$1,680	($6,050)	($14,790)
2-1-39	Wire harness	$4.79	$5,106	$107,168	$112,274	$1,000	$111,274	$350,680
2-1-85	Process tube	$1.04	$3,892	$3,451	$7,343	$340	$7,003	$21,850
2-1-87	Rain shield	$1.45	$11,246	—	$11,246	$3,125	$8,121	$30,614
2-1-89	Sensor support	$.65	$2,444	$2,157	$4,601	$3,000	$1,601	$10,869
2-1-94	Heater wiring harness	$2.27	$1,589	$15,593	$17,182	0	$17,182	$51,583
2-1-96	Copper/aluminum	$.40	$11,862	$11,385	$23,247	0	$23,247	$90,649
2-1-97	Electrical heat kit	$5.87	$4,109	$40,321	$44,430	0	$44,430	$133,290
	Class two proposals							
2-2-2	Extruded drain pan	$2.00	$47,308	$12,000	$59,308	$29,200	$30,108	$154,510
2-2-4	High-strength steel	$13.46	$14,362	$304,021	$318,383	$102,200	$216,183	$852,949
2-2-5	Remove skid	$6.35	$150,202	$38,100	$188,302	$6,200	$182,102	$557,270
2-2-7	One check valve	$4.48	$22,840	$10,061	$32,901	$70,000	($37,099)	$29,129
2-2-13A	Indoor fan motor	$2.74	$2,924	$4,020	$6,944	$1,660	$5,284	$19,456
2-2-13B	IFM mounting bracket	$1.97	$2,102	$8,337	$10,439	0	$10,439	$31,719
2-2-14	I.D. blower and housing	$3.27	$3,489	$7,828	$11,317	$5,460	$5,857	$22,581
2-2-20	Low-voltage block	$.82	$14,335	—	$14,335	$19,800	($5,465)	$23,205
2-2-45	Compressor	$1.95	$2,081	$21,419	$23,500	$4,900	$18,600	$67,578
2-2-61	Alternate filter dryer	$1.14	$5,993	—	$5,993	$6,700	($707)	$11,279
2-2-90	Fan blade	$.97	$28,764	$16,103	$44,867	$16,500	$28,367	$117,883
2-2-91	Compressor tube	$1.40	$41,427	$188,400	$229,827	$3,540	$226,287	$613,830
	Class three proposals							
2-3-17	Works without heater	$.84	$7,135	$27,036	$34,171	$48,750	($14,579)	$53,763
2-3-21	Heat pump control	$31.19	$220,732	—	$220,732	$76,500	$144,232	$585,696
2-3-43	Enhanced fins	$13.46	$318,383	—	$318,383	$125,900	$192,483	$868,861
3-3-73	Programmable test	$2.75	$65,048	$16,500	$81,548	$80,000	$1,548	$187,636
2-3-74	Auto strapping	$.77	$18,213	$4,620	$22,833	$20,000	$2,833	$50,751
2-3-75	No weird angles	$3.00	$70,962	$18,000	$88,962	$46,000	$42,962	$229,655
2-3-92	O.D. fan orientation	$6.28	$109,812	$111,970	$221,782	$179,500	$42,282	$485,846
2-3-95	Defrost sensor	$.77	$2,899	$2,558	$5,457	$2,100	$3,357	$14,271

163

TABLE 11-2. WORKLOAD AND RESOURCE ALLOCATION BY WEEK

	(Number of hours required in week)													(Dollars in week of obligation)		
Wk	Design chief	Design eng'r	Dfting desnr	Dfting detail	Model shop	Test	Tool design	Tool room	VA Team	Eng O/H	Mfg O/H	Pur O/H	Tooling Cost	Material Cost	Capital Cost
1	20	30							50	60		40			
2	20	60							50	20		40			
3	30	120	40						20	10	20	40			
4	20	160	60	40					20	10	60	20			
5	20	160	80	40					20	10	60	20			
6	10	160	120	80	40				20	20	60	20		$1,800	
7	10	160	200	80	40				20	20	60	20			
8	10	160	200	120	40				20	10	80	20			$32,500
9	10	140	220	120	40				20	10	80	20			
10	10	140	220	120	40	40			20	10	80	20		$7,200	
11	10	140	220	120	40	40			20	10	80	20			
12	20	140	220	120	80	40			20	10	80	20			
13	20	120	220	120	80	80			20	10	80	20			
14	20	120	220	120	80	80			20	10	110	20		$2,200	
15	20	120	220	120	80	80			20	10	120	20			
16	20	120	200	120	80	80			20	10	120	20			
17	30	120	200	120	80	80			20	10	120	20			
18	30	120	180	120	40	80			20	10	140	10			
19	30	100	180	120	140	80			20	10	120	20			
20	30	100	180	120	40	80			20	10	120	20			$74,600
21	20	100	160	120	40	80			20	10	120	20			
22	20	80	160	120	40	40			20	10	120	20			
23	20	80	160	120	40	40			20	10	100	20			
24	10	80	160	120	40	40			20	10	100	10			
25	10	80	120	120	40	40			20	10	100	10			

Week												
26	10	80	120	120	60	40	—	—	20	10	100	10
27	10	80	120	120	60	40	—	—	20	10	100	10
28	30	80	120	120	60	40	—	—	20	10	120	20
29	30	80	100	120	60	40	—	—	20	10	120	20
30	20	80	100	100	60	40	—	—	20	10	120	20
31	20	80	100	100	40	40	20	40	20	10	120	20
32	30	100	100	100	40	60	20	80	20	10	120	20
33	30	100	120	100	40	60	80	120	20	10	120	20
34	30	80	120	100	40	60	80	120	20	10	120	10
35	20	80	120	100	20	60	80	120	20	10	120	10
36	20	80	100	100	—	60	80	140	20	10	120	10
37	20	100	100	80	—	60	80	140	20	10	120	10
38	20	100	100	80	—	60	80	140	20	10	180	20
39	20	100	100	80	—	60	80	200	20	10	160	20
40	20	100	100	80	20	40	80	200	20	10	160	20
41	20	100	80	80	20	40	40	200	20	10	160	20
42	10	80	80	80	40	40	40	120	20	10	120	20
43	10	80	80	80	40	40	40	120	20	10	100	20
44	10	80	80	80	40	40	80	60	20	10	100	10
45	10	100	40	40	40	40	80	60	20	10	100	10
46	20	100	40	40	20	40	80	60	20	10	100	10
47	20	100	40	40	—	40	80	40	20	10	100	10
48	10	100	40	40	—	20	80	40	20	10	100	10
49	10	100	40	40	—	—	40	20	20	10	100	10
50	30	80	40	40	—	—	40	20	20	10	100	10
51	30	80	40	40	—	—	40	20	20	10	100	10
52	20	60	40	40	40	40	40	20	20	10	100	10

$8,400 $18,300 $184,500 $11,600

Week one = January 1, 1990

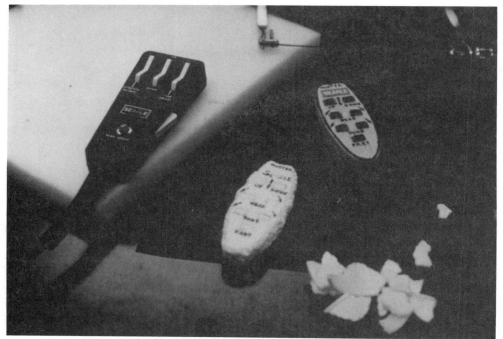

Figure 11–7. Mockups of two stages in the development of a hand control for an X-ray machine were the most effective way to illustrate the team's development of the new concept from the existing control at left.

but the requirement to modify and update the data makes a computer program very desirable. The labor portion of the report (the first 12 significant columns) is simply a summation of all of the labor elements defined in the Gantt charts prepared for each proposal. The dollar-obligation portion of the report (the last three columns) is based on best estimates, by purchasing, tool design, model shop, and industrial engineering, of the cost and lead time required for toolroom material, model shop material, and capital expenditures. It is common for two investment summaries to be prepared: the one shown lists labor hours per week in each of the labor categories; the other, not shown, lists the costs of that labor. Generation of that second summary is simply a matter of multiplying each hours figure by a labor rate, a totally automatic process in a computer-based system. A set of subtotals by month is also often added. This dollar-based data is of particular interest to executive management when making the implementation decision.

The column headings should be adapted to the unique constraints of the organization. Where investments in field service and refurbishing labor are important factors, for instance, include columns to tabulate these investments. Multinational parts procurement sometimes requires a statement of currency conversion assumptions.

The data of Table 11–2 clearly reflect the result of careful labor-load "smoothing." It is common for the first iteration of this chart, being simply a summation of the proposal Gantt charts, to reflect labor requirement curves that are highly discontinuous, suddenly jumping from a low requirement to a high requirement and then back down again. This

Figure 11–8. The team captain unveils a detailed scale model of a newly designed manure spreader; behind both is a poster showing the machine's key features.

is clearly an impractical schedule, since it would require a continuous cycle of rapid hiring and firing.

It is important that the value analysis coordinator perform a smoothing activity on labor-loading before the data is delivered to management at the final presentation. The process requires shifting of the time-bars on each proposal Gantt chart until the obligated labor is balanced with the available labor in each labor category.

The ideal computer program readjusts each proposal Gantt chart element by closing the loop with the time-phased investment summary. In the absence of a computer, manual manipulation of the data is straightforward, though the process commonly requires several iterations.

THE FINAL REPORT

The final presentation is really only the first day of the implementation process. It is essential that a bridge be constructed between the problem-solving activity that characterized the previous steps of the job plan and the critical solution-implementing activity that occupies the remainder of the job plan effort.

This bridge is the written final report.

Copies of the written final report are given to all participants in the value analysis study and to all attendees at the completion of the final presentation to management. It is also supplied to the organization library. It contains all the information necessary to assure maximum probability that the proposals for change will be implemented.

Figure 11–9. *A presenter moves opaque sheets in order to expose data, shown via an overhead projector, in a controlled manner.*

The contents of the written final report typically include the following:

Product description and rationale for value analysis study

Objectives of the study

List of the team members and their areas of specialization

Focus panel or questionnaire data

Function diagram with costs and user/customer attitude allocations

List of value analysis targets

Summary of proposals—by team, by class of proposal, by year, and by investment required (time-phased)

A copy of each value analysis proposal form with associated backup data (typically grouped by team and class of proposal)

Combinex© analysis so that the rationale for proposal selection can be understood.

Each participant leaves the final presentation session with a copy of this written final report. Each is thus equipped to participate in the implementation process.

The champion has accepted responsibility for pursuing the championed proposals until they are either implemented, rejected, or clearly transferred to another person for implementation or investigation. The final report serves as the champion's master checklist for the implementation activities of the follow-up phase.

During the debriefing session after the final presentation, executive management commonly commits to a follow-up procedure to ensure that the results of the value analysis study are not lost. For executive management, then, the written final report serves as a checklist against which to measure the bottom-line effectiveness of the value analysis system.

As in all things, middle management is commonly short on authority and long on responsibility. The written final report serves as the bridge between executive management, from which all authority to act ultimately springs, and the subordinates who investigate and implement value analysis proposals. It can also serve as an information base for some future value analysis team.

In addition to the participants listed above, the written final report serves obvious functions for such groups as the comptroller's office (for audit), industrial engineering, manufacturing engineering, tooling, marketing, field service, and so on (to guide their contributions toward implementation of the value analysis proposals).

Chapter **12**

THE FOLLOW-UP PHASE

The single most important step in the value analysis system is follow-up, to ensure the implementation of all of the team's proposals.

This has also been historically the most poorly handled step.

The focal point of the follow-up process is the value analysis council, chaired by the top operating executive or manager of the organization.

The ideal follow-up integrates the application of the four documents and four elements of structure shown below.

Documents
1. Implementation planning worksheet
2. The value analysis proposal forms
3. The Gantt charts for each proposal
4. The time-phased investment summary

Structure
5. Informed, interactive involvement of the top operating executive
6. Regular, scheduled meetings of the value analysis council
7. An in-house focal point: the value analysis coordinator
8. Funding

DOCUMENTS

The follow-up process actually started with the first team meeting, described in chapter 3 (see page 30). The document prepared at that time is the first one listed here, the implementation planning worksheet.

The worksheet, updated regularly throughout the value analysis job plan, was then used to plan a strategy for the final presentation to management. It remains a key document in the follow-up process.

Two sets of documents that focus on individual implementation effort were published as the body of the written final report distributed at the final presentation: the value analysis proposal forms and their associated Gantt charts. Each proposal form and Gantt chart is a self-contained package that details the implementation effort, the effect on cost and user acceptance, and the essence of the change.

From the management viewpoint, the time-phased investment summary is a critical document. The major question for executive management to resolve in deciding to implement a package of value analysis proposals is, quite simply: "Is it worth making the investment?" The investment summary communicates an uncommonly effective answer to that question, particularly when the individual labor costs are presented in dollars with monthly subtotals. The summary admittedly covers only the financial implications of the change package. Those elements of the package that improve user acceptance, being less quantifiable, are not included in the highly specific and auditable summary; they are included in the individual proposals and in the summaries of the final report.

STRUCTURE

Even with a strong and continuing focus on implementation, and even with proposals that may be obviously brilliant, it is by no means certain that they will be implemented. More is needed than great ideas and well-constructed plans.

Although each organization must clearly fit the follow-up process to its own personality, it is essential to include the four elements of structure listed above.

The informed, interactive involvement of the top operating manager is the most important determinant of the effectiveness of value analysis. In an organization where the top operating manager chooses to delegate the responsibility and authority for the value analysis system to the managers of engineering, manufacturing, or finance, the system is significantly weakened. Value analysis is a horizontal process. It requires a common involvement of all of the separate functional areas of the organization. Only through authority from the top can such horizontal involvement succeed.

Robert H. Brethen is the chief executive officer of Philips Industries of Dayton, Ohio, a *Fortune 500* company that has built one of the most effective value analysis records in history. He has published six "value analysis principles" to guide his multiplant system. The first and most important of these principles is that "top management must be committed and involved." Brethen's level of involvement is best exemplified by his eight years of regular monthly reviews of the status of his value analysis system. Brethen rigorously reviews a monthly report, on a form similar to that shown in Figure A-5 in Appendix A. Best illustrating his commitment is his regular pattern of closing the loop by immediately contacting any division manager whose value analysis performance appears to deviate from the plan.

The second most important element is regular, scheduled meetings of the value analysis council (see chapter 1).

The first such meeting to focus on follow-up is the debriefing session, held immediately following the final presentation to management. Attendees at this debriefing include all members of the value analysis council and the value analysis coordinator.

DEBRIEFING

The singular objective of the debriefing session is to maximize the probability that the just-presented value analysis proposals will be implemented.

The final presentation has appropriately been referred to as the first day of the implementation process. It follows then, that the debriefing is the kickoff to that implementation system.

The agenda for the debriefing will vary greatly depending on the participants, the culture of the organization, and the value analysis history of the organization. The suggested agenda includes the following:

The value analysis council chairman's reaction to the presentation

Reactions by each executive or manager present

Firming of a schedule for regular review and implementation of proposals

Scheduling of future value analysis studies

This agenda is based on the following assumptions:

This is the organization's first value analysis study.

A top operating executive or manager was responsible for instituting the value analysis system.

Executive management, though having attended a briefing, has had no intimate involvement in establishing or monitoring the study.

The organization has suffered the not-uncommon experience of unfulfilled promises in experimenting with new programs. It is thus a bit skeptical of the promises that the value analysis teams have just made.

A full-time value analysis coordinator has been appointed and is present.

A sketch is presented at the outset of the debriefing session as the key discussion document for planning a formal effort to implement the proposals (Fig. 12–1). This sketch clearly identifies the value analysis council as the focal point for follow-up activities. It also illustrates the integrating influence of the full-time value analysis coordinator.

Agenda

The debriefing is guided through its agenda by the value analysis coordinator. The first and most obvious item is to request comments from the chairman of the value analysis council. These comments will set the tone for the debriefing session. If the chairman has been properly advised of the objectives, he or she may elect to lead the entire session.

It is important that all executives and managers present be offered an opportunity to express their reactions to the final presentation. This invariably reveals differences in perception, which must be reckoned with during the follow-up process.

A specific schedule must be set for future follow-up actions. This should include a

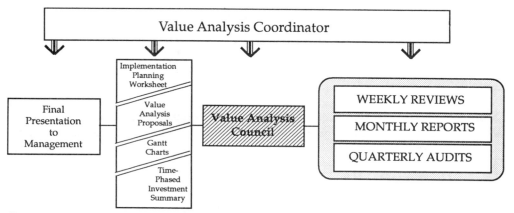

Figure 12–1. This block diagram illustrates all of the fundamental elements of the follow-up and implementation process.

weekly schedule for value analysis council reviews of proposal status, as well as specific deadlines for other required implementation activity.

POLICY AND PROCEDURE

A standardized policy and procedure is detailed in Appendix A. It was formulated for a multidivisional company with a central value analysis system but with responsibility for the cost improvement and value analysis systems delegated to division management. It covers all of the elements of a complete cost-improvement system, including value analysis, cost reduction, delivery cost reduction, and operations or methods improvement.

It is recommended that a copy of Appendix A be given to the organization's comptroller, with the request that it be modified to fit the constraints of the organization and then issued by executive management.

Chapter 13

CASE STUDIES

This chapter is a collection of case studies of value analysis efforts. They were not selected at random. Each illustrates an element of the value analysis system in a way that should trigger in the reader a flash of recognition of a similar situation from a past or present design effort.

In each case study, highly pertinent elements of the value analysis system are emphasized and detailed, rather than those elements that were less influential in reaching the conclusions of the case study.

While the "before" and the "after" are part of each case study, the reader is urged to look beyond the "present" and "proposed" blocks on the value analysis proposal forms shown, and instead reflect on the overall system by which the team and the champion reached their conclusions.

Do not search through the case studies for specific solutions to copy; search through them for the appropriate system used to find the solutions. The U.S. government has indeed developed a data bank of previously developed value analysis solutions, but the government environment is primarily oriented to cost reduction and often operates through second-guessing. Under such conditions, a data bank of solutions may be of use. In modern value analysis, however, the focus is on the system. The system *is* the solution.

The before-and-after format as used in this chapter is a powerful technique for proving a point or illustrating a process—perhaps too powerful, since it can mislead.

A very effective example of the use of this format in the early history of value analysis is the widely publicized "dog-and-pony show" of Secretary of Defense Robert S. McNamara. This presentation was displayed annually at that highest of all levels, the White House press conference, to prove to the entire country that McNamara's value engineering program, which he called "eliminate goldplating," was saving specific dollars in specific defense programs. The Society of American Value Engineers, the American Ordnance Association, and others prepared a poster-based series of value engineering examples, each of which consisted of a picture of a "before" and one of an "after," together with a summary of the defense dollars saved.

This poster display went on tour to many U.S. cities to spread the word further that a formal program of value engineering was expanding the effectiveness of defense dollars.

This standard format, with the "old way" on the left and the "new way" on the right, has proven, over many years and many thousands of uses, to be an effective shorthand method of demonstrating a point. This chapter emphasizes the system used to achieve the results, but the before-and-after technique is used as a supplement to each of the case studies to illustrate a number of examples of successful value analysis.

The author's caution about this technique is based partly on an early experience at the Laboratory for Electronics in Boston, where he had just accepted a job as value analysis manager. In order to strengthen his presentations in a value analysis workshop, he borrowed some before-and-after posters from an associate who led the value analysis effort at the Porter Cable Company in Syracuse, New York. On the left of each cardboard poster was a piece of "before" hardware from a Porter Cable power tool. The "after" hardware was on the right. As an experiment, he removed all of the hardware from the posters and reversed the order, putting the "before" on the right and the "after" on the left. During the many hours of the value analysis study, not one of the participants raised a question. They simply accepted that the "before" was on the left.

This experiment was intended to illustrate to the participants that it is not possible to judge from simple observation whether a product has good value. It additionally proved to the author, and to the several dozen participants, that the technique can be insidious. Have a healthy skepticism when evaluating such an example. Don't be misled by positioning.

Each case study includes the following:

A definition of the product

The system by which the solution was developed

A concise statement of the lesson learned

In most cases, a virtual copy of the completed value analysis proposal form (each including the "before" and "after" in the form of a sketch or statement under the headings "Present" and "Proposed."

Most of the case studies are unexpurgated. Some are modified for one of three reasons: to avoid revealing proprietary information; to avoid revealing confidential costs; or to satisfy requests by the sponsoring organizations that their activities not be published.

The first seven case studies were selected from actual proposals presented by the team studying the "Theme Thread" product as described in chapter 3 (see Fig. 3–5). An index to all the case studies, including lessons learned and key points, constitutes Appendix C.

1. Reduction of Check Valves
The Product: The "Theme Thread" heat pump (see Fig. 3–5)
Lesson Learned: Wild card works

A team of five General Corporation decision makers from engineering, advanced manufacturing engineering, marketing, finance, and purchasing applied the value analysis system to their 4-ton heat pump.

Their FAST diagram defined 46 independent functions, one of which—under the primary function of "enhance product" —was "modulate capacity." Function-cost data was then allocated to the diagram, revealing that "modulate capacity" accounted for 7 percent of the total product cost. The data from the user focus panel, when allocated to the diagram, clustered in the "modulate capacity" area, with several important likes and two significant dislikes. The team therefore identified it as a value analysis target and posted the function "modulate capacity" at the top of a flip chart. They recorded 180 words and phrases, plus a few entries that could be characterized as ideas, on four different sheets in a no-judgment brainstorming session. A portion of the 180 entries is shown in Figure 13–1.

During the synthesis phase, the team saw a tantalizing challenge in the entry "no check valves." A team member volunteered, "We should be able to delete some of the five check valves, but we'd have to carefully relocate the capillaries."

From that unlikely initial trigger phrase came value analysis proposal number 2-2-7 (Fig. 13–2). Three-year net savings are conservatively predicted to be $29,000.

MODULATE CAPACITY

SWITCH	THERMAL
VALVE	ORIFICE
WIRES	CAPILLARY
QUEUE	SENSOR
SEQUENCE	DIVERT FLOW
THERMOSTAT	USE CENTRAL
SONAR	DATA PROCESSOR
VIBRATION	WITH SOLID STATE
RADIO CONTROL	SENSORS
CLOSED LOOP	LIQUIFY
AUDIO	DIRECT GAS
TAPE RECORDER	ELIMINATE TUBING
MEASURE NEED	NO CHECK VALVES
USE CENTRAL	NO FITTINGS
PROCESSOR	MANUAL CONTROL

Figure 13–1. A portion of a flip chart shows brainstorming words and phrases on the value analysis target "modulate capacity."

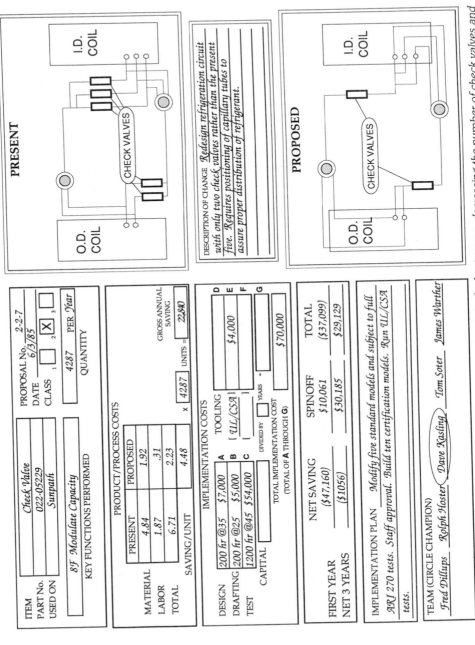

Figure 13–2. *Value analysis proposal 2-2-7 suggests a change in the heat pump, decreasing the number of check valves and probably saving at least $29,000 over three years.*

The success of this team in tolerating ambiguity by honestly attempting to apply the summarily ridiculous flip-chart entry "no check valves" is simply another illustration of the power of the wild card technique as described in chapter 8.

2. The 90-Degree Concept
The Product: The "Theme Thread" heat pump
Lesson Learned: Unconstrained creativity pays off

The same team that proposed and implemented case study 1 had one member who was particularly sensitive to the machine shop setup and forming labor in this heat pump with so many angled components. His job was advanced manufacturing engineering, an assignment that brought him into daily contact with design engineering in keeping with the General Corporation's major effort to improve the manufacturability of new products.

In this capacity he often suggested ways to minimize the setup and labor in forming sheet metal covers, dividers, supports, and brackets. He deeply appreciated how much extra effort was required when any angle except 90 degrees was called out. He seldom presumed, however, to suggest that the design engineers make dramatic changes in the basic geometry of a unit. Where a unit contained many acute and obtuse angles, his basic assumption was that there were technical reasons that components had to be angled to each other to improve refrigerating or heating efficiency.

In the environment of the value analysis study team, however, he functioned as a coequal member of a design team. The function orientation and the unconstrained creativity of the brainstorming session set the stage for his comment, in the synthesis phase, that "if everything was at right angles, a lot of labor could be saved." He agreed to champion an investigation, in cooperation with the design engineer, with the objective of changing all hardware to a 90-degree orientation.

The result was proposal 2-3-75 (Fig. 13–3).

The cross-fertilization of ideas and concepts, which resulted here in nearly a quarter of a million dollars of probable savings within three years, is a major side benefit of the team-based value analysis study. The relationships and the mind-set that inevitably develop in each of the team members invariably increase the likelihood of such multidisciplinary cooperation in product improvement.

3. Enhanced Fins
The Product: The "Theme Thread" heat pump
Lesson Learned: Old but good ideas never die

The project engineer on the refrigeration system was a team member. He had urged, for several years, that enhancement of the fins should be considered to improve efficiency and permit a reduction in the number of fins per coil. During the function-cost allocation session on the second day of the study, he raised this concept as a result of the microallocation of the elements of the coil. He was urged to record the concept in the Idea Bank. His entry read, "Enhance fins and reduce number of fins."

When the team entered the synthesis phase in the middle of the fourth day of team sessions, the project engineer volunteered to become the champion of the concept. Some tests were performed and the labor and material costs were estimated with the support of the team member from advanced manufacturing engineering. This resulted in value analysis proposal number 2-3-43 (Fig. 13–4).

This proposal exemplifies the capability of a value analysis study to sweep up all of

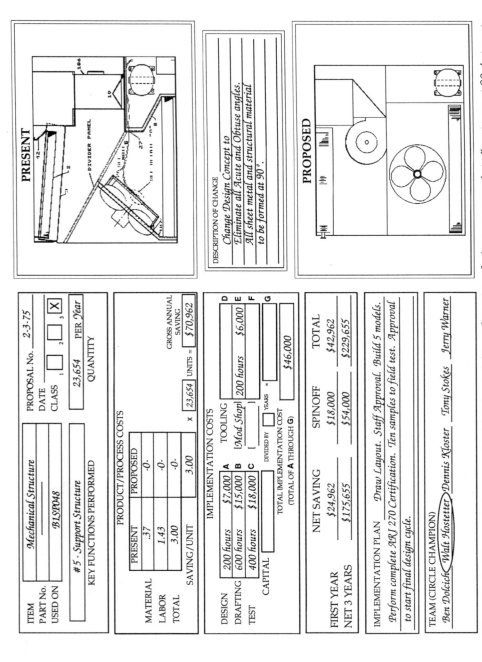

Figure 13–3. Value analysis proposal 2-3-75 changed the configuration of a heat pump so that all parts are at 90 degrees to others.

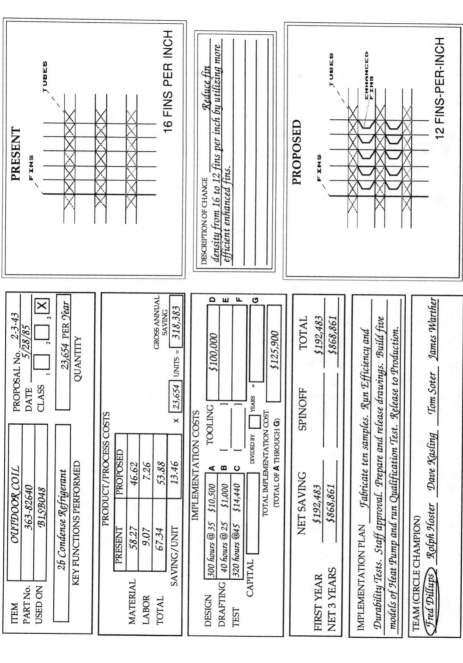

Figure 13–4. Value analysis proposal 2-3-43 enhances the fins and reduces fin density in a heat exchanger in order to improve efficiency and reduce cost.

the possible product improvements that have idled about in the organization but have not been implemented, for a great variety of reasons. It is common for such concepts to constitute 20 to 60 percent of the total proposals included in the team's final report.

4. Keyholed and Flanged Side Panels
The Product: The "Theme Thread" heat pump
Lesson Learned: The value analysis target can be based solely on worth

In identifying a value analysis target to focus the problem-solving effort of a value analysis study, the team analyzes both the cost and the worth of each function. Most targets are at least partially based on high function cost. In this case, the target was based solely on some worth data collected from users/customers in the prestudy focus panel. Of the 82 responses from the panel, 2 likes and 1 dislike were allocated by the team to function 5B2, "enhance access." These three entries in the focus panel report are shown in Figure 13–5.

The creativity session, which focused on the function "enhance access," led to a number of different championed concepts. Six of these survived the development phase and were included in the team's final report. Proposal number 2-1-26 (Fig. 13–6) is typical of three of these six, in that it does not propose a cost reduction.

Indeed, it requires a 25-cent increase in the cost of every heat pump to change all side panel holes to keyholes and form a 90-degree lip on each side panel.

It is worth noting that the other three championed concepts triggered by the "enhance access" target, unlike the first group of three, required no cost increase. All three responded to the user need to improve access, and all three saved money in the process.

5. Delete Fan Motor Bracket Clamp
The Product: The "Theme Thread" heat pump
Lesson Learned: Experts and vendors have good ideas—use them

A major theme of the value analysis process is "Look beyond your organization." During the synthesis phase, a group of outside experts and vendors is brought into the session to act as sounding boards and idea provokers for team champions.

During a vendor interplay session, the heat pump team demonstrated to the fan motor vendor the way the motor mounts into the structure. He immediately suggested that the present bolted cylindrical clamp be dropped, and that his company weld the three mounting brackets to the motor shell. A fast telephone call to his company

	LIKES (Features or Characteristics)															MODE
13 Service Panels have "Zeus" fasteners	6	8	6	8	8	10	10	6	8	3	9	9	9	10	10 5	8R7
20 Minimizes Field Labor	8	8	8	9	8	9	9	9	10	10	9	1	10	10	10 8	9R9
	DISLIKES (Faults or Complaints)															
20 Field Disassembly Tricky	8	10	9	10	8	10	3	9	10	9	10	10	9	10	9 8	10R7

Figure 13–5. An excerpt from a focus panel report shows two likes and one dislike, all three reflecting user attitudes toward field repair.

PRESENT

SCREW MUST BE COMPLETELY REMOVED WITH A SCREWDRIVER

SHARP EDGE HARD ON FINGERS

HOLE STRIPS OUT AFTER 4 OR 5 REMOVALS

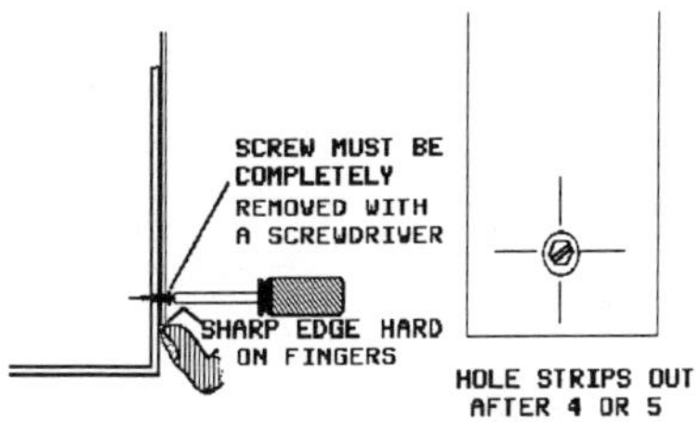

DESCRIPTION OF CHANGE On all side panels, change the clearance hole to a keyhole slot, permitting removal of panel without removing screw. Form bottom of panels at 90 degrees by 3/8 inch to prevent damage to the fingertip of the maintenance person.

PROPOSED

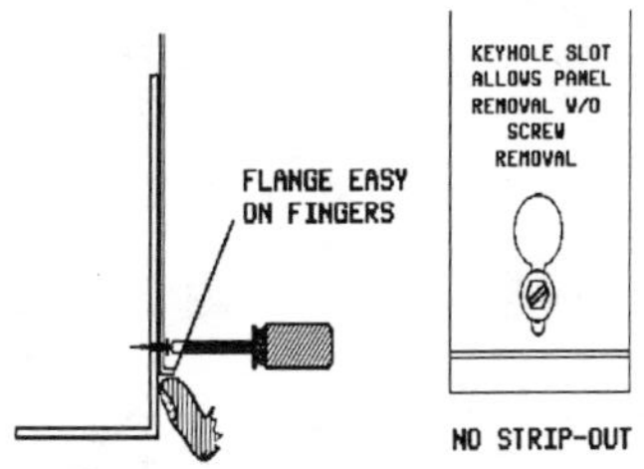

FLANGE EASY ON FINGERS

KEYHOLE SLOT ALLOWS PANEL REMOVAL W/O SCREW REMOVAL

NO STRIP-OUT

ITEM	Side Panels		PROPOSAL No.	2-1-26
PART No.	Various		DATE	5/28/85
USED ON	Sunpath-Pathfinder		CLASS	[X] 1 [ ] 2 [ ] 3

KEY FUNCTIONS PERFORMED   Reduce Downtime

17,482 PER Year   QUANTITY

PRODUCT/PROCESS COSTS

	PRESENT	PROPOSED
MATERIAL	$4.60	$4.67
LABOR	$2.17	$2.35
TOTAL	$6.77	$7.02
SAVING/UNIT		($0.25)

x 17,482 UNITS = GROSS ANNUAL SAVING ($4,370)

IMPLEMENTATION COSTS

DESIGN	8 Hours @35	$280	A	
DRAFTING	24 Hours @25	$600	B	TOOLING $500 D
TEST	4 Hours @45	$180	C	[Model 1] 4 Hours @30 $120 E
				[ ] F
CAPITAL				G

DIVIDED BY [ ] YEARS =

TOTAL IMPLEMENTATION COST (TOTAL OF A THROUGH G) $1,680

	NET SAVING	SPINOFF	TOTAL
FIRST YEAR	($6,050)		(6,050)
NET 3 YEARS	($18,150)		($18,150)

IMPLEMENTATION PLAN   Draw new delineations. Tool for forming operation. Issue ECN. Issue PCN. Change Manual. Immediate action

TEAM (CIRCLE CHAMPION)
Fred Dillups   Rolph Hoster   Dave Kasling   Tom Soter   James Warther

Figure 13–6. Value analysis proposal 2-1-26 changes heat pump side panels to ease disassembly, prevent degradation, and minimize injury.

183

confirmed that this operation could be performed at no additional cost. The present purchased brackets, at a cost of 61 cents for three, will be shipped to the motor vendor for welding in place. Figure 13–7 shows a net three-year saving, on the unit under study and on several other models, of over $31,000.

This change proposal resulted only indirectly from the unique function mind-set of value analysis. It is typical, however, of a significant portion of most teams' proposals. It resulted from a receptive and motivated team asking an expert or vendor for his or her best ideas on how to improve the product. The proven effectiveness of this procedure is dramatically better than the typical results of an institutionalized Vendor Day with its tables of tagged parts and its stacks of formal blueprints.

6. Low-Voltage Block Relocation
The Product: The "Theme Thread" heat pump
Lesson Learned: Improving cost and worth concurrently

The function "reduce downtime" was identified by the team as a value analysis target. Its cost was only 0.7 percent of the total, but the participants in a user focus panel clearly classified the function as important to them. The panel also identified two significant complaints about downtime caused by the difficulty in disconnecting such items as the heater kit.

The mind-set of the team members in attacking this target was "Damn the extra cost; just reduce the downtime!" When they tackled the problem in the synthesis phase, they discovered that by moving the low-voltage terminal block from the heater kit to the basic unit, the main field control cable could be permanently installed. The addition of a four-pin connector to the heater kit permitted rapid removal and replacement of the kit.

When the champion was verifying the costs and feasibility of the concept, he discovered that the in-house on-line testing and hookup were both greatly simplified; therefore, the change resulted in a new cost saving of $23,000 for the first three years, in addition to greatly simplifying field installation of the heater kit (Fig. 13–8).

This double-barreled result—reducing cost while improving customer satisfaction—is not an exception. It is the nearly universal result of the dynamic of a value analysis team study.

7. Field Conversion, Side to Down
The Product: The "Theme Thread" heat pump
Lesson Learned: A life-cycle cost advantage drives the decision to implement

The value analysis team, as well as the marketing and engineering management of the General Corporation, was predisposed to accept that a new design concept was justified. Field support had long complained that competition was more easily converted from "side" to "down" input and output.

The desirability of this redesign effort was confirmed when the team analyzed the data from a user focus panel comprising eight dealers and distributors and eight key General Corporation marketing, engineering, and quality managers. The pertinent data are summarized below:

Function: 5A2, "permit modification"

Cost: $9.56 (0.1 percent of the total)

PRESENT

FABRICATED
BRACKET
TO BE
CLAMPED TO MOTOR

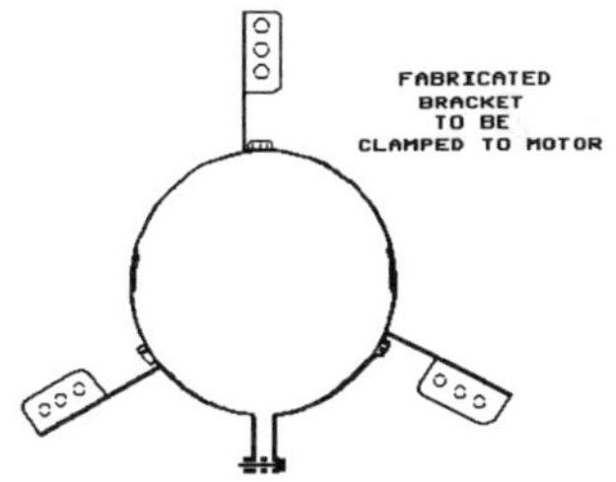

DESCRIPTION OF CHANGE *Eliminate the in-house fabricated motor mounting bracket. Motor vendor will weld our brackets to the motor shell at no charge.*

PROPOSED

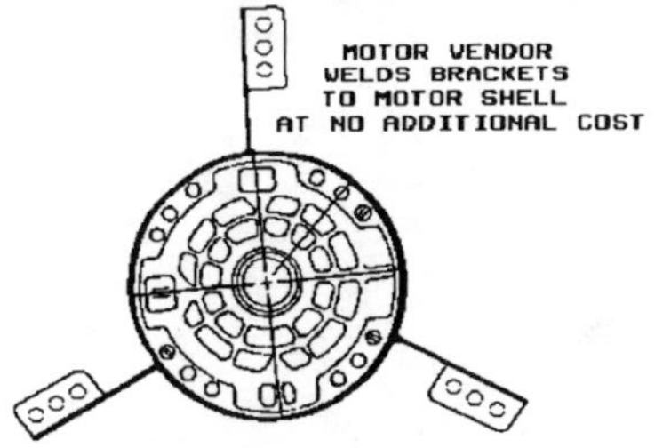

MOTOR VENDOR
WELDS BRACKETS
TO MOTOR SHELL
AT NO ADDITIONAL COST

ITEM	IFM Mounting Bracket
PART No.	363-73300C001
USED ON	B1SP048

PROPOSAL No. 2-2-13
DATE 5/30/85
CLASS 1 [ ] 2 [X] 3 [ ]

QUANTITY 1,067 PER Year

KEY FUNCTIONS PERFORMED 4G2 - Protect Components

PRODUCT/PROCESS COSTS

	PRESENT	PROPOSED
MATERIAL	$.89	$.61
LABOR	$1.69	$.61
TOTAL	$2.58	$1.97
SAVING/UNIT		

x 1,067 UNITS = GROSS ANNUAL SAVING $2,102

IMPLEMENTATION COSTS

		TOOLING	
DESIGN	A	D	
DRAFTING	B [ ]	E [ ]	
TEST	C [ ]	F	
CAPITAL		G	

DIVIDED BY [ ] YEARS =

TOTAL IMPLEMENTATION COST (TOTAL OF A THROUGH G) $ -0-

	NET SAVING	SPINOFF	TOTAL
FIRST YEAR	$2,102	$8,337	$10,439
NET 3 YEARS	$6,206	$25,011	$31,217

IMPLEMENTATION PLAN *Ship 300 brackets to vendor. Verify integrity during standard incoming test. Change standard Motor P/O.*

TEAM (CIRCLE CHAMPION)
Fred Dillups   Ralph Hoster   Dave Kasling   Tom Soter   James Warther

Figure 13–7. Value analysis proposal 2-2-13 suggests deleting a fan mounting clamping assembly, and having the motor vendor attach brackets to the motor shell.

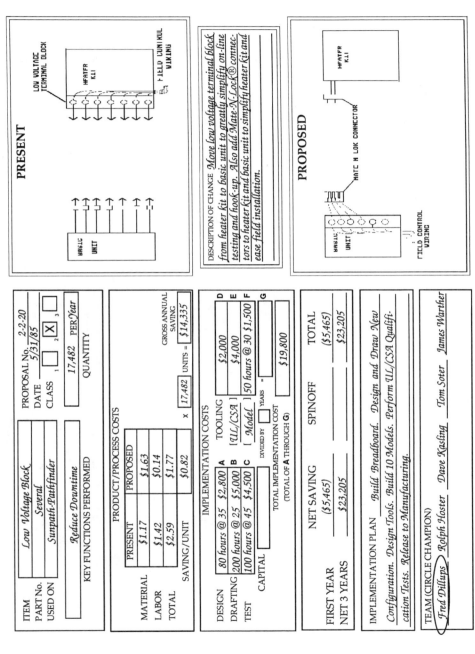

Figure 13–8. *Value analysis proposal 2-2-20 moves the heater kit to the basic heat pump unit, simplifying testing, hookup, and field assembly.*

User focus panel response: A wish-list item was expressed by the panel as follows: "Easily convertible from side output to down." The vote on the desirability of the feature was 10R5. (Fifteen panelists rated it at 8, 9, or 10.)

During the synthesis phase, the project engineer quickly volunteered to champion the major redesign. His feasibility and cost review revealed that the new configuration represented only a $20 (2 percent) increase in product cost. He further estimated that the reduction in the carrying cost of inventory, both in the plant and at distributor locations, would more than offset the added cost. The field labor saving was calculated to be $775,000 per year, based on the estimate that 30 percent of the present units in the 3- to 5-ton range require conversion.

Note that the team members chose to restructure the value analysis proposal form to fit the constraints of this unique project. They first calculated the gross annual saving, based on the data available from field support. They then calculated the unit saving by dividing the gross annual saving by the predicted annual quantity. This restructuring of the data permitted them to present an integrated summary of all proposals to management in the final presentation. The value analysis proposal is shown in Figure 13–9.

8. Use Low-Cost Spark Plugs
The Product: 1,200-horsepower gasoline engine
Lesson Learned: Value analysis targets trigger insight

Three high-powered teams applied the value analysis process to the entire $102,000 engine. Each team focused its function-cost activity on the segment of the engine with which it was most familiar. They all function-allocated the 31 likes and the 31 dislikes that were previously expressed in a one-day user focus panel.

The teams concluded that the costs of six of the functions were not well matched to the needs and wants of the users (Table 13–1).

Triggered by these six functions, the teams identified over 100 specific changes for investigation. Of these, 56 survived and were published in the team's final report.

In many cases, the link between the value analysis target and the final proposal is tenuous. At first sight, for example, this proposal to substitute low-cost spark plugs during in-house testing would seem to have little relationship to any of the six functions in Table 13–1. The fact is that those targets did lead to the proposal. This is a demonstration of the powerful dynamic of the value analysis process. The mind-setting that started with the prestudy reading assignment and carried through the rigorous, step-by-step function analysis loaded the minds of the team members with a massive set of valid data. The triggering of this data through focusing on value analysis targets is the source of most of the dramatic accomplishments of a modern value analysis team.

TABLE 13–1. VALUE ANALYSIS TARGETS FOR AN EIGHT-CYLINDER ENGINE

No.	Function	Present cost
6C	Minimize maintenance	$ 70.07
4A3	Lubricate surfaces	$2,068.32
4B1	Maintain balance	$1,344.54
4D	Minimize defects	$ 463.39
5A1	Fasten components	$2,049.07
6D	Increase efficiency	$ 887.01
		$6,882.40
		(64% of total)

PRESENT

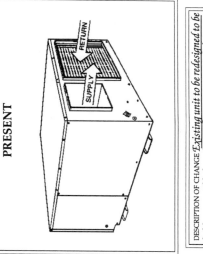

SUPPLY
RETURN

PROPOSED

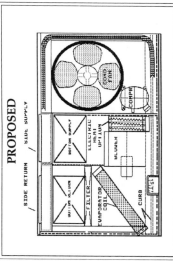

SIDE RETURN / SIDE SUPPLY

UNIT ON RETURN | BOTTOM SUPPLY | ELECTRIC HEAT OPTION
FILTER | COND FAN | COMPR
EVAPORATOR COIL | CURB | BLOWER

DESCRIPTION OF CHANGE *Existing unit to be redesigned to be field-convertible from bottom input-output to side input-output. There is a $20 cost adder in the new design. In addition to the significant field labor saving, there is probably a greater than $20 offset through lower inventory carrying cost for both the field and factory.*

ITEM	*Field Conversion*	PROPOSAL No.	*2-3-67*
PART No.	*DNA*	DATE	*6/11/85*
USED ON	*B1SP048*	CLASS	☐₁ ☐₂ ☐₃ ☒

KEY FUNCTIONS PERFORMED *592 Permit Modification*

$23654 \times .3 = 7096$ PER *Year*

QUANTITY

(Assume 30% of 3 to 5 Ton units will use plenum)

PRODUCT/PROCESS COSTS

PRESENT	*7,096 units X $176 (current plenum cost + labor*	$1,248,896
PROPOSED	*23,654 units X $20 (extra cost/unit to convert)*	$ 473,080

GROSS ANNUAL SAVING | **$775,816** divided by | **7096** | =

UNIT SAVING | **$109.33**

IMPLEMENTATION COSTS

					TOOLING		
DESIGN	*100 hours @ 35*	$3,500	**A**			$50,000	**D**
DRAFTING	*80 hours @ 25*	$2,000	**B**				**E**
TEST	*320 hours @ 45*	14,400	**C**				**F**
CAPITAL							**G**

DIVIDED BY ☐ YEARS =

TOTAL IMPLEMENTATION COST | $69,900 *(in-house costs)* | **G**

(TOTAL OF **A** THROUGH **G**)

	NET SAVING	SPINOFF	TOTAL
FIRST YEAR	$705,916	- 0 -	$705,916
NET 3 YEARS	$2,353,454	- 0 -	$2,353,454

IMPLEMENTATION PLAN *Redesign and Draw. Build 10 Models. Perform complete ARJ Qualification Test. Release to Manufacturing. Design and Build Tools. Prepare and Implement Marketing Plan.*

TEAM (CIRCLE CHAMPION)

Fred Dillups *Rolph Hoster* *Dave Kasling* *Tom Soter* *James Warther*

Figure 13–9. *Value analysis proposal 2-3-67 suggests redesigning the configuration of the heat pump to be field-convertible from side to down input-output.*

The engine was equipped with eight super-durable, long-life spark plugs, which cost nearly ten times as much as the standard 89-cent model from an auto store. Two sets were shipped with the engine. The set in the cylinders is subjected to the standard run-in test at the factory and mating-up tests at the packager. The second set is supplied so that when the engine is put into service, it can start out with fresh plugs.

The team noted a $149.76 cost for spark plugs during the cost-function allocation. When the team member from finance asked what was so exotic about these plugs, he was told that these were special long-life plugs. The brief discussion that followed triggered proposal 2-1-8 (Fig. 13–10). Standard 89-cent plugs have the same low failure rate in the first few hours of use as the $9.36 specials, so, the finance team member wanted to know, "Why not use the cheap plugs to check out the engine, and put the long-life plugs in when the engine is put into service?"

This change was implemented immediately, with little paperwork, and on all models of engines, more than doubling the saving.

9. Buy Pushrod Adjusting Screws Outside
The Product: 1,200-horsepower diesel engine
Lesson Learned: Make versus buy

The manager of manufacturing engineering was a value analysis team member. During the function-cost analysis, he was startled to see that each engine required $178 worth of pushrod adjusting screws. They were being machined in-house at $11.14 each, and 16 were needed per engine. He made an entry in the idea bank and, during the synthesis phase, agreed to champion an investigation into what seemed like an excessively expensive piece of threaded steel.

While he checked the factory routing, he asked purchasing to find an outside source for cost comparison. A screw-machine house found the part to be a natural and quoted a price per screw of $1.73, an 81 percent reduction.

This proposal uncovers a classic case of creep, as described by Robert L. Dixon, PhD, CPA, in an article under that name in the July 1953 issue of *The Journal of Accountancy*. Dixon presents a persuasive argument that many business failures result from the cumulative action of bringing in-house numerous activities that could be more effectively produced outside. The champion never deduced the temporary circumstances that had caused this pushrod adjusting screw to be tooled, routed, and scheduled for in-house manufacture. The reasons really didn't matter. The $150-per-engine penalty on this small item was a signal that a broad, critical, make versus buy review was overdue. This rather extreme example (Fig. 13–11) became the justification for a general make versus buy review.

10. Fiberglass Doors
The Product: Continuous-flow production oven
Lesson Learned: You can always make a good product better

A not-uncommon constraint on a value analysis redesign is the requirement that any change must not be obvious. This limitation is most common when the current product is highly accepted in the marketplace and any change might affect that acceptance.

The unique proposal shown in Figure 13–12 was developed under such a constraint. The bolted-on access doors on the side of the heat chamber were cast-iron monsters weighing 40 and 60 pounds. The process of placing the doors in position and inserting

PRESENT

M14 X 0.7/ THREAD

SIXTEEN SUPPLIED PER ENGINE

($7.01 EACH)

2.30 · 1/2 RLM · .125 DIA WIRE · .87 · .59 · .500 USP · REF .22 · GAP .018 · M18 X 1.5 THD · RECT SECTION .062 X .125

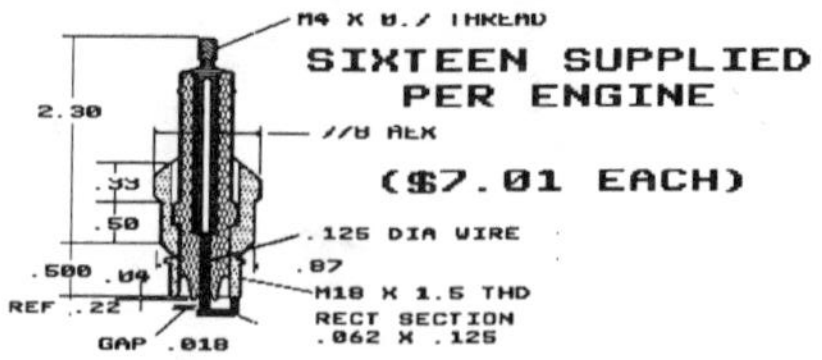

Two sets of plugs are supplied with each engine, since the set in the engine has been used in endurance testing in the factory and by the packager. When the engine is ready to be put into service, the old set is discarded and a fresh set is installed.

DESCRIPTION OF CHANGE *Supply engines with low cost spark plugs in cylinders. Supply an extra set of long-life plugs to be installed when engine is put into service.*

PROPOSED

Supply 8 low-cost ($.89) spark plugs with new engine for use while performing in-house endurance and packaging tests. The failure rate of low-cost plugs during the relatively limited period of these tests will be no higher than the failure rate of the long-life, high-cost plugs.

Ship engine to packager with the low-cost plugs in the cylinders. Ship eight long-life plugs with the engine to be put in place when engine is placed in service.

ITEM	Spark Plugs
PART NO.	17A1211
USED ON	Model 8422

PROPOSAL No. 2-1-8
DATE 12/10/85
CLASS 1 [X] 2 [ ] 3 [ ]
84 PER Year QUANTITY

KEY FUNCTIONS PERFORMED: *3C Ignite Charge*

PRODUCT/PROCESS COSTS

	PRESENT	PROPOSED
MATERIAL	74.88	7.12
LABOR		
TOTAL	74.88	7.12
SAVING/UNIT	67.76	

x 84 UNITS = $5,691

GROSS ANNUAL SAVING $5,691

IMPLEMENTATION COSTS

A	TOOLING	
B		
C		
DESIGN		D
DRAFTING		E
TEST		F
CAPITAL		G

TOTAL IMPLEMENTATION COST (TOTAL OF **A** THROUGH **G**)

DIVIDED BY [ ] YEARS = - 0 -

	NET SAVING	SPINOFF	TOTAL
FIRST YEAR	$5,691	$8,262	$13,953
NET 3 YEARS	$17,073	$24,786	$41,859

IMPLEMENTATION PLAN *Issue ECN. Revise BOM. Procure new short life plug. Manufacturing Engineering will implement shop floor change.*

TEAM (CIRCLE CHAMPION)
Tom Fader   Ted Mover   Jim Bachbert   John McFee   (Ed Jackson)
Ben Echberg

Figure 13–10. *Value analysis proposal 2-1-8 changes $7.01 plugs to $.89 plugs for engine checkout and run-in.*

PRESENT

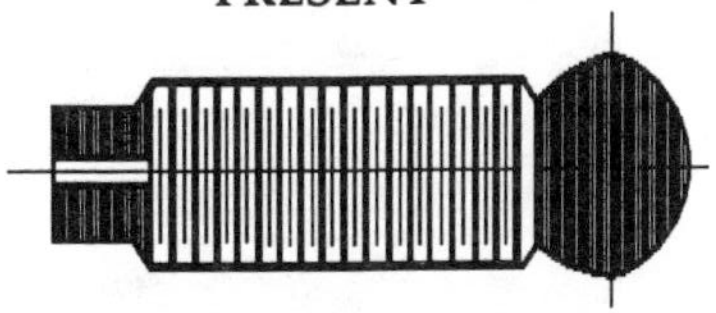

MADE IN-HOUSE @ $11.14
(Current Annual Volume 3919)

DESCRIPTION OF CHANGE *Change from Make to Buy to take advantage of Vendor's specialized equipment and experience.*

PROPOSED

BUY OUTSIDE @ $1.73
From I&P Centerless Grinding Co.

(PART UNCHANGED)

ITEM	*Pushrod Adjusting Screw*
PART No.	*15B2014*
USED ON	*All Model 8 Engines*

PROPOSAL No. *2-1-10*
DATE *1/10/86*
CLASS 1 [X] 2 [ ] 3 [ ]

124 PER *Year*
QUANTITY

Control Sequencing
KEY FUNCTIONS PERFORMED

PRODUCT/PROCESS COSTS

	PRESENT	PROPOSED
MATERIAL	$ 3.84	$27.68
LABOR	$174.40	
TOTAL	$178.24	$27.68
SAVING/UNIT	$150.56	

$150.56 × 124 UNITS = $18,669

GROSS ANNUAL SAVING
$18,669

IMPLEMENTATION COSTS

A		D	
B	TOOLING	E	
C		F	
	DIVIDED BY [ ] YEARS =	G	
CAPITAL			

TOTAL IMPLEMENTATION COST
(TOTAL OF **A** THROUGH **G**) - 0 -

	NET SAVING	SPINOFF	TOTAL
FIRST YEAR	$ 18,669	$25,021	$43,690
NET 3 YEARS	$56,009	$75,063	$131,070

IMPLEMENTATION PLAN *Order sample lot. Inspect. Buy production quantities.*

TEAM (CIRCLE CHAMPION)
(George Fall) Gunter Ferd Alice McClay Todd Joss Ed Blaser
Bruce Leeder

Figure 13–11. Value analysis proposal 2-1-10 changes pushrod adjusting screws from make to buy, a big reduction from in-house production costs.

191

PRESENT

CAST IRON

60 Lbs.

40 Lbs.

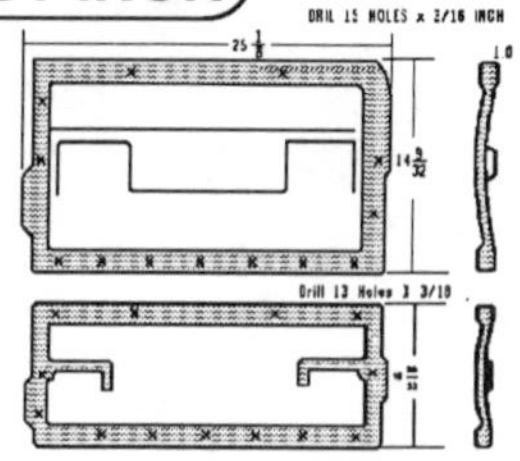

DRILL 11 HOLES x 3/16 INCH
.18 INCH THICK
Drill 12 holes 3 7/8

HEAVY - BULKY
APPEARS STRONG

Both covers use non-reversible gaskets - leaking problems.

DESCRIPTION OF CHANGE *Change material only (configuration and appearance remain identical) from Cast Iron to Fibreglass-reinforced High Temperature Plastic. Mounting present door requires two men. The fibreglass door is an easy job for one man.*

PROPOSED

FIBREGLASS

LIGHT - BULKY
APPEARS STRONG

Weight 8 pounds

Is actually stronger than Cast Iron.

Resin is colored grey and *looks like* Cast Iron.

O-Ring groove molded in - fewer leaks and re-usable.

Light. Can be handled easily by one man.

ITEM	*Cover Doors*
PART No.	*134B2244*
USED ON	*CF-32 Oven*

PROPOSAL No. 1-2-3
DATE 12/10/87
CLASS 1 [ ] 2 [X] 3 [ ]

KEY FUNCTIONS PERFORMED *Appear Strong*

57 PER *Year* QUANTITY

PRODUCT/PROCESS COSTS

	PRESENT	PROPOSED
MATERIAL	$320	$110
LABOR	$257	$110
TOTAL	$577	$110
SAVING/UNIT		$477

x 57 UNITS = $27,189 GROSS ANNUAL SAVING

IMPLEMENTATION COSTS

					TOOLING	
DESIGN	8 hours	@22	$176	A	$13,000	D
DRAFTING	32 hours	@18	$576	B		E
TEST	40 hours	@20	$800	C		F
CAPITAL					$14,552	G

DIVIDED BY [ ] YEARS =

TOTAL IMPLEMENTATION COST
(TOTAL OF **A** THROUGH **G**)

	NET SAVING	SPINOFF	TOTAL
FIRST YEAR	$12,637	$46,000	$58,637
NET 3 YEARS	$67,015	$138,000	$205,015

IMPLEMENTATION PLAN *Have a sample cover molded and test in lab, then in field.*

TEAM (CIRCLE CHAMPION) *Catherine Moore John Moroni James Fong Rick Black Jerry Lockworth*

Figure 13–12. Value analysis proposal 1-2-3 changes the cover doors from bulky, expensive cast iron to fiberglass-reinforced plastic colored to appear like cast iron.

the bolts required two strong men. In an attempt to determine the customer attitude toward heavy doors, one of the company participants in a focus panel suggested the following "fault" in the oven: "Nonstructural doors, covers, and hatches are massive and heavy-duty."

This dislike was voted by the 17-man panel as a "4" on the 1-to-10 scale of seriousness. In other words, the massiveness, with all of its handling problems, is regarded by the customer as being an advantage, probably because weight is being equated with strength.

The team, alerted by the dislike, launched into a discussion of alternative design approaches to the access doors. Value analysis proposal 1-2-3 changes the material to iron gray, high-temperature, fiberglass-reinforced plastic—which is actually stronger. The cost reduction of 81 percent is dramatic. Acceptance in the field has yet to be proven, but since the gray color looks like cast iron, it is likely that the customer will notice no difference, at least until the repairman is required to remove a door.

11. Cast-in Lubrication Header
The Product: Horizontal milling machine
Lesson Learned: Improve value by changing outmoded company policy

The company operated an in-house foundry, which permitted great flexibility and control over development, design, and scheduling of cast components. Costs were higher than typical quotations from outside vendors, but the conventional wisdom was that the advantages of control were worth the extra cost.

As in so many cases of vertical integration, the captive foundry was not required to compete in the open market, and therefore it failed to expand its capabilities when new casting methods became available outside, unless the new method was required by a present product.

It was a fairly efficient foundry when producing the standard castings that were typical of past and present company products. It was not, however, capable of such new methods as no-bake molding.

The value analysis team, during its function-cost analysis, had jotted in its Idea Bank a suggestion that the fabricated lubrication header assembly be deleted and replaced by a cast-in gallery as an integral part of the machine bed. The team members knew that such a change would require no-bake molding, and that therefore the machine bed would have to be produced at an outside foundry. They also knew that such an action was essentially forbidden by company policy.

They recorded the idea nevertheless, because the value analysis system had developed in them a collective mind-set that could tolerate ambiguity; that is, they could permit themselves, at least temporarily, to pretend that even those things that are clearly impossible should be considered.

During the synthesis phase, the project engineer, who was a team member, agreed to champion the concept. With the support of a senior buyer and the methods supervisor, who were members of other teams in the same workshop study, he rigorously defined the cost of the new concept. It reduced the lubrication header cost by $302, or 77 percent. This was too good not to be presented to management for reconsideration of company policy. The change was presented as value analysis proposal 3-2-5 (Fig. 13–13).

The net saving, after a $4,000 tooling cost and $6,300 in implementation labor, was $101,000 in the first three years. Management was presented with a clear question:

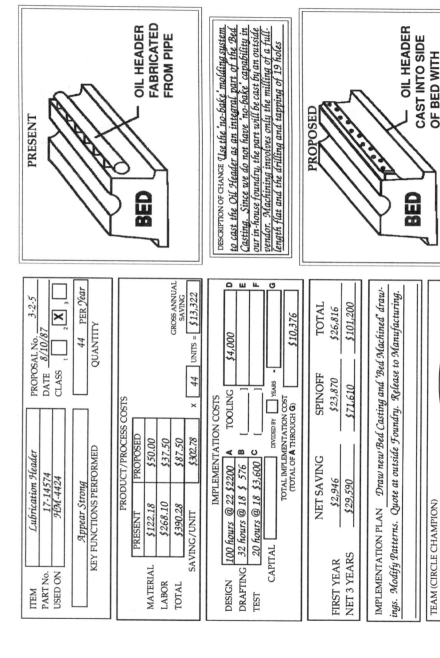

Figure 13–13. Value analysis proposal 3-2-5 modified the header casting to incorporate the oil header into the bed—for a three-year net saving of $101,000.

How does the loss of some direct foundry labor balance with the average $34,000-per-year cost saving?

12. Bearing Temperature System
The Product: 1,200-horsepower stationary prime mover
Lesson Learned: Increased MTBF improves value

The user focus panel evaluated a complaint or dislike: "oil leaks." The eight company participants rated this dislike at 5 (with a voting range of 5 to 9). The nine actual users rated it at 9 (with a voting range of 3 to 9). With the clear difference of perception between the company and the users, a discussion ensued that clarified the attitude of the users. Leaks were apparently a tolerated fault until they became ubiquitous. At that point, the fault became nontolerated. The value analysis team tackled the question rather directly, identifying the function "minimize leaks" as a value analysis target.

During the creativity and synthesis phases, this targeted function triggered a discussion of field complaints, which focused the team's attention on several leakage areas including the network of tubing used to interconnect the nine heat-activated pressure relief valves mounted on the main bearing caps. This, in turn, led the manager of manufacturing engineering, a team member, to champion a critical look at the entire bearing temperature system.

The resulting proposal (Fig. 13–14) requires a $3,000 initial investment but saves $5,700 per year on the engine under study. The selling point for the proposal, however, was not its moderate cost saving potential; it was the total elimination of the oil leakage potential of the bearing temperature system.

13. Unshielded Ignition System
The Product: 1,200-horsepower engine
Lesson Learned: Eliminate unwanted feature and reduce cost

The General Corporation engine is well regarded in the market. Packagers, who sell the engine as an integrated element in engine-compressor systems, find that its dependability and other key features make it an ideal choice for remote, heavy duty-cycle applications. The company has included as standard features a number of subsystems that are not provided by competition. One of these is a totally shielded ignition system, which prevents any possible spark in the ignition primary from igniting any flammable gas in the vicinity of the engine. The company advertises this feature and assumes that sales benefit by including it in the standard engine.

The user focus panel revealed that the users were not impressed; it also revealed that company sales, marketing, field engineering, and quality managers were surprisingly not impressed either (Fig. 13–15).

The first eight respondents were members of company management; their votes ranged from 2 to 5. The nine actual users voted, with one exception, 1 to 3. Clearly, nobody cared much for the shielded system, a feature that adds $700 to each engine so equipped. The shielded ignition system was an obvious value analysis target.

The project engineer was a member of the team. As champion of the concept, he estimated that the change would require only $1,520 for design and drafting. Gross annual saving was $54,000, with a $253,000 net saving for the first three years. The resulting value analysis proposal suggested changing the ignition system from shielded to unshielded (Fig. 13–16).

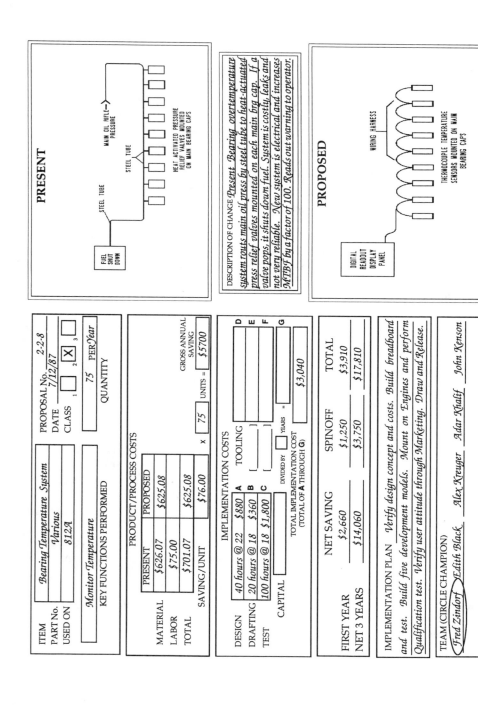

Figure 13–14. *Value analysis proposal 2-2-8 greatly simplified the bearing overtemperature system and greatly improved reliability, completely eliminating potential oil leakage.*

	LIKES (Features or Characteristics)																	
	1	2	3	4	5	6	7	8	9	10	11	12	13	14	15	16	17	**MODE**
28 Shielded Primary Ignition System	3	3	2	4	5	2	3	5	2	3	3	3	1	2	2	1	5	3R4

Figure 13–15. This excerpt of the focus panel results reveals that the shielded ignition system is not needed by most users.

No price reduction for the engine is appropriate. For the few customers who may request shielding, it will be supplied at extra cost.

14. Water-Cooled Exhaust Manifold
The Product: 1,200-horsepower engine
Lesson Learned: Make it safe, improve value

The engine is supplied with an insulated air-cooled manifold. The high temperatures require stainless steel for the manifold and the wastegate piping. The user focus panel revealed that the high surface temperatures were a serious customer concern. Fault (dislike) number 28 was considered serious by a significant number of the voters (Fig. 13–17). The panel discussed this problem at length, with the users referring to it as a safety hazard and quoting instances in which the high temperature has started fires.

Note that the four people voting 10 were all actual users. Company participants were more muted in their concern about the difficulty of working around a hot manifold.

The project engineer accepted this project as its champion and performed a detailed cost estimate, which indicated that the user concern could be eliminated while reducing the cost of the manifold/wastegate by $5,000 per engine! The major reason for this windfall is the substitution of carbon steel for stainless steel, since the temperature environment of both is substantially reduced.

Value analysis proposal 3-2-17 (Fig. 13–18) indicated a probable saving of nearly $4 million in the first three years while improving the product for the customer!

15. Air Motor Redesign
The Product: Actuator for bus and subway doors
Lesson Learned: Eliminate faults, give customers what they want, and improve price and market share

This value analysis study was performed in the early days of the career of Thomas F. Cook, CVS, FSAVE. Its conclusions and his association with Thomas J. Snodgrass, the father of user-oriented value analysis, led him later to apply the user-oriented approach throughout his many years of successful value analysis. The bus door motor (Fig. 13–19) was a major product of Cook's employer, a division of a major, multidivisional U.S. corporation. Its market share, however, was only 9 percent and dropping.

The responsible general manager was told by corporate management that unless he increased his sales and his market share within six months, the product line would be closed down. This proved to be an effective motivator, and the general manager authorized a value analysis team workshop study.

Cook led the study, using the function analysis process developed by Snodgrass, the founder of Value Standards, Inc. of Chicago. This was the early 1960s, and the FAST

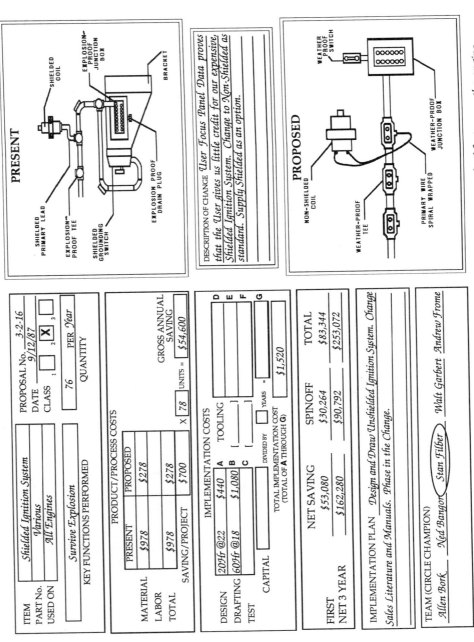

Figure 13–16. Value analysis proposal 3-2-16 deletes the shielded ignition system as a standard feature on the engine.

	DISLIKES (Faults or Complaints)																		
	1	2	3	4	5	6	7	8	9	10	11	12	13	14	15	16	17	**MODE**
28 Hot Manifolds (Hard to Work Around)	4	8	4	5	8	5	8	8	10	5	6	3	9	10	10	10	6	8R7

Figure 13–17. An excerpt from focus panel data shows how much the voters were concerned about high temperatures around the manifold.

diagram system of function analysis would not be conceived by Charles W. Bytheway at Univac until the mid-1960s.

Cook led his team through the function definition process until the salesman on the team suddenly rebelled. According to Cook, he said, "Wait a minute! You're all assuming that you know the reasons a person buys these units. *I* don't even know the reasons, and I have to sell this stuff!" The remainder of this discourse illustrates what Cook claims is his realization of the critical importance of user data in a value analysis study.

The salesman was asked how the team could increase its knowledge of the customers' needs and wants. The salesman said, "Let's ask them!" This fit well with the Snodgrass concept of value analysis, which required the interview of actual users. As a result, key members of the team traveled to Indianapolis and Boston to ask various users what turned them on—and off—about the door motors (Fig. 13–20). The users consulted included the following:

- bus builders
- subway car builders
- transit system maintenance mechanics
- transit authorities
- riders

The group returned from the trip with the following list of needs and wants:

- pneumatic operation
- fast opening and closing
- cushioned opening and closing
- speed adjustment in both directions
- nonslamming after release of held door
- various positions (shaft up, down, left, right)
- minimum air loss
- minimum weight

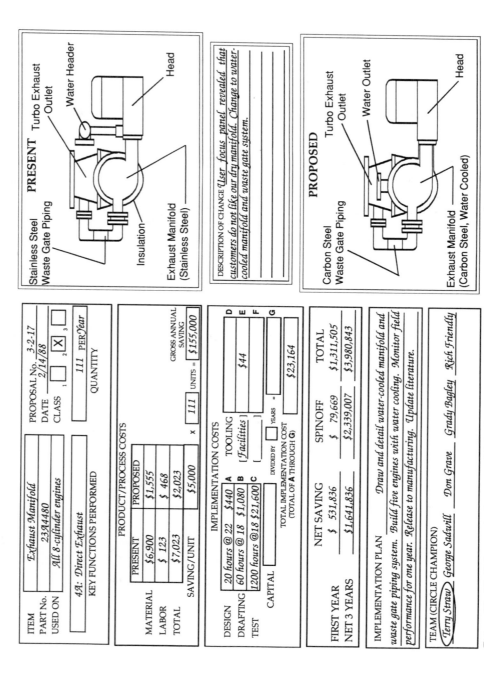

PRESENT

Turbo Exhaust Outlet

Water Header

Head

Stainless Steel Waste Gate Piping

Insulation

Exhaust Manifold (Stainless Steel)

DESCRIPTION OF CHANGE *User focus panel revealed that customers do not like our dry manifold. Change to water-cooled manifold and waste gate system.*

PROPOSED

Turbo Exhaust Outlet

Water Outlet

Head

Carbon Steel Waste Gate Piping

Exhaust Manifold (Carbon Steel, Water Cooled)

ITEM	*Exhaust Manifold*	PROPOSAL No. *3-2-17*
PART No.	23A4480	DATE *2/14/88*
USED ON	*All 8-cylinder engines*	CLASS 1 ☐ 2 ☒ 3 ☐

4A: Direct Exhaust
KEY FUNCTIONS PERFORMED

111 PER/Year
QUANTITY

PRODUCT/PROCESS COSTS

	PRESENT	PROPOSED
MATERIAL	$6,900	$1,555
LABOR	$ 123	$ 468
TOTAL	$7,023	$2,023
SAVING/UNIT		$5,000

GROSS ANNUAL SAVING

× 111 UNITS = $155,000

IMPLEMENTATION COSTS

DESIGN	20 hours @ 22	$440	A	TOOLING	$44	D
DRAFTING	60 hours @ 18	$1,080	B	[Facilities]		E
TEST	1200 hours @18	$21,600	C	[ ]		F
CAPITAL				DIVIDED BY ☐ YEARS =	$23,164	G

TOTAL IMPLEMENTATION COST
(TOTAL OF A THROUGH G)

	NET SAVING	SPINOFF	TOTAL
FIRST YEAR	$ 531,836	$ 79,669	$1,311,505
NET 3 YEARS	$1,641,836	$2,339,007	$3,980,843

IMPLEMENTATION PLAN *Draw and detail water-cooled manifold and waste gate piping system. Build five engines with water cooling. Monitor field performance for one year. Release to manufacturing. Update literature.*

TEAM (CIRCLE CHAMPION)
Terry Straw   George Sadwill   Don Grave   Grady Bagley   Rich Friendly

Figure 13–18. Value analysis proposal 3-2-17 changed the manifold and waste gate to a water-cooled system.

Figure 13–19. This bus door motor, shown before redesign, led Thomas F. Cook to user-oriented value analysis.

Figure 13–20. Subway car builders are interviewed inside a subway car about the motor for the doors.

- long service life
- low initial cost
- low-cost replacement parts
- convenient to replace cup seals
- few parts to stock

These items of information were not rated in importance. They were not even evaluated on the basis of whether they were being expressed by a prime buying influence. They were simply thrown into the pot and stirred by the team.

But this team was singularly motivated. The members' jobs hung in the balance. They had talked face-to-face with actual users. They were using that most powerful of all industrial problem-solving systems: value analysis. As a result, the team returned a series of eight key redesign proposals, which promised to reduce the cost by 50 percent as well as measurably increase user acceptance. These eight proposals are described below.

Proposal 1: The mounting system comprised split clamshell clamps (Fig. 13–21). Each of the two clamps is made up of two die-cast arms with a mounting platform and a pair of bolts with elastic stop nuts. The concept of the original design was that such a flexible arrangement would permit the bus or subway car manufacturer to mount the door motor in any orientation throughout 360 degrees.

Figure 13–21. Proposal 1: The original clamshell clamps for holding the motor in place were excessively flexible, and the elastic stop nuts proved ineffective.

The user data proved that the car manufacturers did not require this flexibility. They had only four options for mounting: up, down, left, or right. In addition, the users complained that even with the elastic stop nuts, the vibration on the roads or rails eventually jarred the units into rotating on their axis, causing the actuating arms to jam suddenly.

The redesign (Fig. 13–22) took advantage of four through bolts to mount a simple angle iron. It can be shifted easily to any of four orientations and has the additional advantage that, even if the lightly stressed nut/lock washers loosen, the unit merely starts to rattle. It is essentially impossible for the loosened unit to jam, and the rattling will attract the attention of the mechanic during regular maintenance.

The angle bracket redesign saved 34 percent of the function cost and measurably increased product acceptance, both by minimizing weight and by reducing the number and complexity of repair parts that must be kept in inventory.

Proposal 2: The next redesign focused on the speed control. The design concept for the body of the door motor used a large- and a small-diameter steel tube with two sand-cast and machined, threaded end-caps, plus a center section that was also a machined sand casting, threaded to the tubes on both ends (Fig. 13–23).

The larger of the two sand-cast, threaded end-caps contained a pipe-tapped hole, into which a speed control assembly was mounted. This was a brass casting with spring-loaded valves and adjusting screws to control the opening and closing speed of the doors.

The team redesigned the motor to use two sets of four through bolts each in place of

Figure 13–22. Proposal 1: The new angle bracket mount reduced cost, weight, and inventory problems, increasing user acceptance.

Figure 13–23. Proposal 2: The original sand-cast assembly relied on a separate assembly to control the door's opening and closing speed.

the threaded end caps (Fig. 13–24). In addition, the large end cap was changed from a machined sand casting to a net-shape die casting; the casting is cored and machined to receive the speed-adjusting spring-loaded valves.

The saving in function cost was 79 percent. Weight is reduced. The potential for air leakage is reduced. Replacement parts in the event of failure are reduced to the spring-loaded valves only. The through-bolted assembly greatly decreases the effort required to change the piston seal in regular maintenance.

Proposal 3: The team then turned its attention to the small end cap and cushion control. The sand-cast small end cap (Fig. 13–25) was threaded to the body of the door motor and contained a separate machined orifice, which controlled the speed of operation of the piston. The through bolt concept permits a die-cast housing to replace the threaded sand casting. The speed orifice is as-cast.

The cost reduction is 55 percent. Weight is reduced. Parts inventory reduction benefits the field stockroom.

Proposal 4: The cylinder liner was replaced. A brass cylinder was threaded into the large-end steel cylinder to provide an acceptable sealing surface for the operating piston. The steel tube is replaced with a section of unthreaded, DOM aluminum tubing, which has an acceptably smooth bore so that no brass liner is required.

Figure 13–24. Proposal 2: The new design is cored and machined for the valves.

Figure 13–25. Proposal 3: The new design of the small end cap also relied on through bolts and die-cast housing.

This change reduced cost by 45 percent. It also contributed to the ease of replacing piston seals in the field. An additional advantage is the greatly improved service life of the aluminum cylinder.

Proposal 5: The central structure of the door motor was sand-cast with each end threaded to accept the small and the large steel tubes. It had a threaded vent hole. The new net-shape die-cast central body requires no threading. The vent hole is cast in place.

The saving in cost was 73 percent, with the added advantage that weight was reduced.

Proposal 6: The piston was machined aluminum with a peripheral groove to carry a felt wiper, which was loaded with oil to improve the wiper-to-cylinder seal and minimize wear. The new piston is molded from Delrin® (Fig. 13–26). No additional lubrication is required, eliminating the felt wiper and its associated oil.

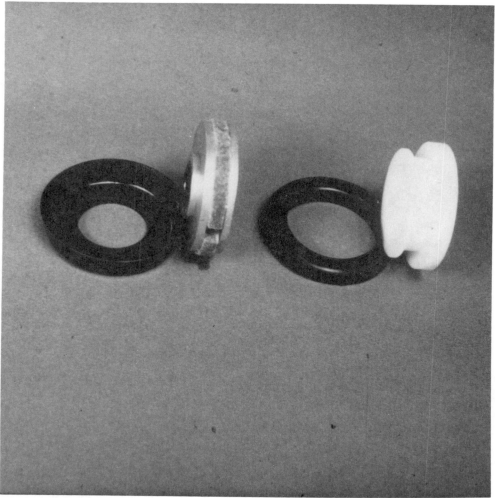

Figure 13–26. Proposal 6: The value analysis team replaced the aluminum piston, left, with one molded from Delrin to eliminate the need for additional lubrication.

Saving in cost was 42 percent. Additionally, air leakage was reduced, piston life was greatly increased, and the effort required to change piston seals was minimized. An additional advantage is the reduced number and cost of replacement parts required in field stockrooms.

Proposal 7: The cushion plunger and guide received the next redesign. When the car door approaches its open or closed limit, it is necessary to damp the action of the door opener to prevent slamming. This action was controlled by a six-part plunger assembly, operating in a deep-drilled hole in the rack gear (Fig. 13–27). The new design eliminates the expensive gun-drilling operation, substituting a spring-floated disc for the plunger shaft (Fig. 13–28).

The new assembly has only four parts and costs 80 percent less than the original. Its life is predicted to be orders of magnitude greater than the solid-shaft version. It reduces weight as well as the number and cost of required field replacement parts.

Proposal 8: Miscellaneous improvements led to a 38 percent cost reduction, through a 20 percent drop in assembly time, the elimination of nonfunctional paint, and the use of a roll pin on the output pinion gear.

The redesigned door motor (Fig. 13–29) represents a 51 percent cost reduction.

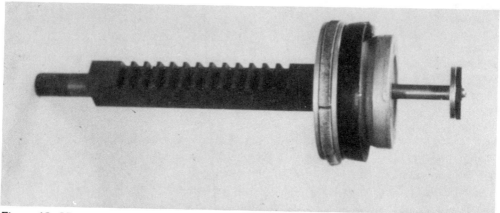

Figure 13–27. Proposal 7: The original cushion plunger and guide had a six-part assembly and an expensive gun-drilled hole in the rack gear.

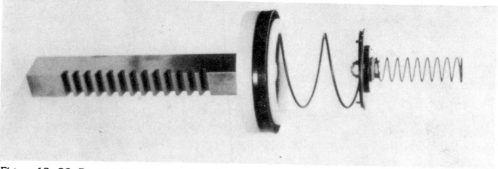

Figure 13–28. Proposal 7: The redesigned assembly, with a spring-floated disc, is more reliable and much less expensive.

Figure 13-29. *As a result of the value analysis team study, the redesigned door operating motor (compare with Fig. 13-19) is less expensive and more acceptable to users.*

The most significant result of the value analysis team redesign was the elimination of the faults or dislikes that were described by the users, and with it the significant improvement of the door motor in areas that the users defined as important to them. The ultimate result of these improvements was a dramatic increase in sales. Three years after the completion of the value analysis redesign, the company's market share had increased from 9 percent to 36 percent, despite a 15 percent increase in price.

16. Multiply Bulkheads
The Product: Catenary anchor-leg-moored (CALM) buoy, Imodco Corporation
Lesson Learned: Wild card works

The product is anchored offshore and used to carry a pipeline to shore for loading and unloading the cargo of supertankers (Fig. 13-30). The fitting on the top of the buoy that attaches to the hoses leading to the bow of the ship has a rotating joint. The body of the buoy does not rotate and is held in location by six radiating anchor chains, which fall to six moorings in catenary curves. Buoys are built in shipyards as near to their ultimate mooring place as possible. The company's production rate was only two per year, due to a weakness in world demand and strong and rising competition.

A team was set up in a hotel room far removed from the nearest shipyard. This violates the fundamental rule that a value analysis team must meet in an area where it can "lay hands on" the product under study. There were two reasons for this compromise: (1) the expense of transporting the team members and value analysis consultants (most buoys are built overseas, most recently in Indonesia); and (2) no buoys were

Figure 13–30. An oil tanker is attached to a catenary anchor leg moored (CALM) buoy, which required redesign to survive accidents better.

starting the construction phase in the immediate future. The team made do with detailed blueprints and a scale model. The team was an ideal mix of the decision makers from the five key areas: design, sales, planning, finance, and field management. The project engineer was a team member.

The team created an excellent FAST diagram and pinpointed 9 of the 41 independent functions as value analysis targets, based on an in-house user focus panel and a rigorous microallocation of the $2 million cost of the buoy.

One of the targeted functions was "survive collision" with a function cost of $44,250. Most of these costs were for a set of six radial bulkheads, which divided the hull into six "unsinkable" compartments (Fig. 13–31). The reason for this structure was the requirement that the buoy remain afloat and that the rotation of the top fixture not be constrained even if the hull is punctured through contact with the bow of a ship. Calculations indicated that the flooding of one compartment would still allow rotation, but the flooding of two compartments would cause the buoy to tilt to the degree that it would not perform properly. A modification to the basic design had been made when a concern was raised: "What would happen if the collision occurred in the vicinity of a bulkhead? Wouldn't that flood the two adjacent compartments?" The solution was to fill every other compartment with plastic foam. This was a problem, because the equipment for producing the foam must be transported from the United States, and the foaming had to be skillfully performed. When conditions were not perfect, the foam had to be chipped out and replaced. All in all, the foam was a problem.

The team headed a flip chart with the function "survive collision" and loaded the sheet with words. One of the team members said, "Eliminate bulkheads." The next said, "More bulkheads." This caused some merriment, since the present six bulkheads seemed like quite enough.

During the synthesis phase, teams are urged to try to make use of every entry on the flip chart, no matter how ridiculous. When the team reached the "more bulkheads" phrase, the project engineer suddenly jumped up and sketched a 12-bulkhead version. This eliminated the foaming, with its attendant cost and problems; ridiculous as it had seemed, the project engineer volunteered to champion the addition of six more bulkheads!

His feasibility and cost study between sessions eventually resulted in a value analysis proposal (Fig. 13–32).

17. Coal-Mining Cutter Head
The Product: Self-propelled underground coal miner
Lesson Learned: Function-cost analysis focuses the value problem

Four value analysis teams segmented the massive 62-ton machine (Fig. 13–33) into four areas of focus. Team 1 worked on the cutter head, team 2 on the gather/convey elements, team 3 on the tractor/convey elements, and team 4 on the electric/hydraulics.

A powerful user focus panel of 12 customers and company managers gathered for six hours to identify their likes and dislikes. Their viewpoint of the so-called coreless cutting feature of the machine was expressed in like number 7 (Fig. 13–34).

Discussion before and after the vote made it clear that the coreless cutting feature pleased the customer—but all competitors delivered the same feature, each using a different method. The function-cost allocation process revealed that this company's design was excessively expensive, since it made use of canted cutting drums: Each of

Figure 13–31. A CALM buoy with six compartments is under construction in a shipyard.

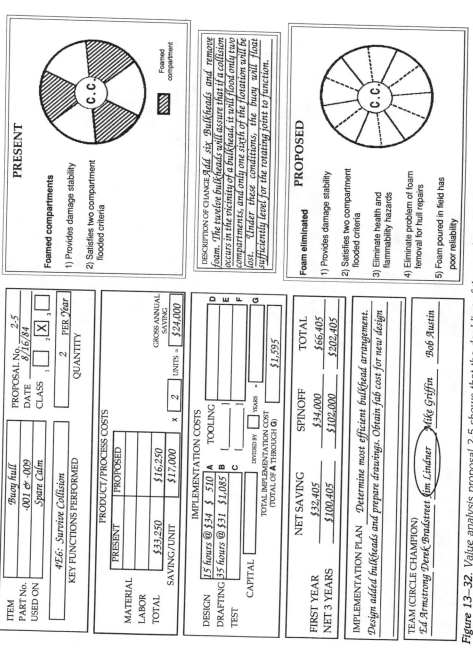

PRESENT

Foamed compartments

1) Provides damage stability

2) Satisfies two compartment flooded criteria

▨ Foamed compartment

DESCRIPTION OF CHANGE *Add six bulkheads and remove foam. The twelve bulkheads will assure that if a collision occurs in the vicinity of a bulkhead, it will flood only two compartments, and only one sixth of the flotation will be lost. Under these conditions, the buoy will float sufficiently level for the rotating joint to function.*

PROPOSED

Foam eliminated

1) Provides damage stability

2) Satisfies two compartment flooded criteria

3) Eliminate health and flammability hazards

4) Eliminate problem of foam removal for hull repairs

5) Foam poured in field has poor reliability

ITEM *Buoy hull*
PART No. *-001 & -009*
USED ON *Spare Calm*

PROPOSAL No. *2-5*
DATE *8/16/84*
CLASS 1 [ ] 2 [X] 3 [ ]

4E6: Survive Collision
KEY FUNCTIONS PERFORMED

2 PER *Year*
QUANTITY

PRODUCT/PROCESS COSTS

	PRESENT	PROPOSED
MATERIAL		
LABOR		$16,250
TOTAL	$33,250	$17,000
SAVING/UNIT		$16,250

× 2 UNITS = $24,000

GROSS ANNUAL SAVING $24,000

IMPLEMENTATION COSTS

				TOOLING		
DESIGN	*15 hours @ $34*	$510	**A**		**D**	
DRAFTING	*35 hours @ $31*	$1,085	**B**		**E**	
TEST			**C**		**F**	
CAPITAL					**G**	

DIVIDED BY [ ] YEARS =

TOTAL IMPLEMENTATION COST $1,595
(TOTAL OF **A** THROUGH **G**)

	NET SAVING	SPINOFF	TOTAL
FIRST YEAR	$32,405	$34,000	$66,405
NET 3 YEARS	$100,405	$102,000	$202,405

IMPLEMENTATION PLAN *Determine most efficient bulkhead arrangement. Design added bulkheads and prepare drawings. Obtain fab cost for new design*

TEAM (CIRCLE CHAMPION)
Ed Armstrong Derek Bradstreet (*Jim Lindner*) *Mike Griffin   Bob Austin*

Figure 13–32. *Value analysis proposal 2-5 shows that the doubling of the number of bulkheads in the CALM buoy actually lowers the cost by $17,000.*

Figure 13–33. The 62-ton underground drum miner, used in coal mines, underwent a four-team value analysis study.

	LIKES (Features or Characteristics)									
	1	2	3	4	5	6	7	8	9	**MODE**
28 Coreless Cutting	8	8	8	8	6	5	8	8	9	**8R4**

Figure 13–34. User focus panel data shows that the users regarded coreless cutting as a very desirable feature.

the two outer drums is inclined forward so that the carbide cutting teeth overlap, removing the core—the portion of the coal seam that would otherwise be left on the wall, preventing the machine from moving forward into the seam. This required two $7,000 heavy-duty, constant-velocity joints, each of which must be isolated from the destructive underground environment. The company regarded its achievement of an inherently coreless drum miner as an important selling point. It developed, however, that customers required core removal but did not care how the machine achieved it.

The project engineer championed the principle of a rotating, reciprocating, cam-operated plate, which is positioned between the center drum and each of the outer drums. It will chew up the core and its cost is a fraction of the U-joint cost. He built a cardboard model (Fig. 13–35). The model was an exhibit in the final presentation to management. The value analysis proposal is shown in Figure 13–36.

Figure 13–35. The cardboard model of a cam-operated cutter was shown at the final presentation.

18. Ceiling Diffuser
The Product: Ceiling-mounted air diffuser
Lesson Learned: Informal user data guides the team

Case study 18 is taken from the extensive records of Philips Industries, a *Fortune 500* company headquartered in Dayton, Ohio. When Thomas Cook Associates introduced it to value analysis in 1980, it was less than one-third its current size. Robert Brethen, the chief executive officer, gives major credit for this growth to his value analysis–focused cost-improvement system.

Most organizations that have established long-term value analysis systems maintain a strong focus on user-need fulfillment. In the Philips System, the user focus is less formal. Its value analysis workshops do not collect user data through questionnaires or focus panels. They do, however, apply user-oriented FAST diagramming, which directs the attention of the value analysis team toward what are called the supporting functions. These are the functions that equate to product quality: "assure dependability," "assure convenience," "enhance product," and "please senses."

A team was formed at Philips's Titus Division to value analyze a ceiling diffuser, a device that fits into the ceiling outputs of a heating and air-conditioning system. The diffuser evens the air flow into the room while harmonizing with the room's decor.

The value analysis team held informal discussions with a number of users/customers to determine their likes and dislikes about the ceiling diffuser. The team then created and costed a FAST diagram of the $10.58 diffuser. This revealed several areas in which function cost failed to match the team's opinion of the users' needs and wants.

These targets are shown in Table 13–2, together with the design changes that the

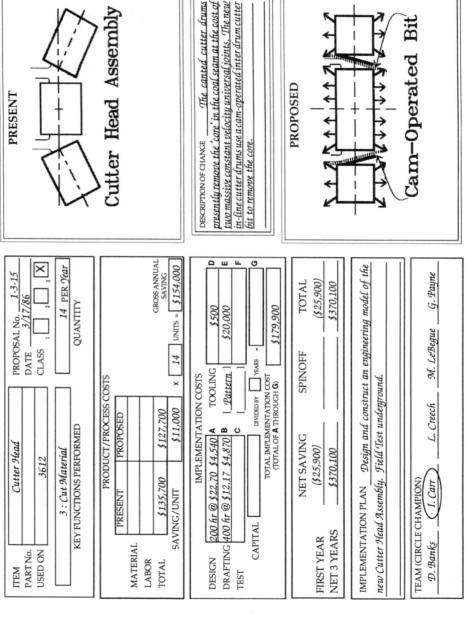

Figure 13–36. *Value analysis proposal 1-3-15 shows the new cam-operated configuration, coupled with the elimination of the angled drums.*

TABLE 13-2. CEILING DIFFUSER VALUE ANALYSIS

Function	Cost	Design Change
Style product	$1.34 (13%)	Delete the center cone.
Minimize housekeeping	$0.05 (0.5%)	Users complained about ceiling smudging caused by air output. The team reshaped the remaining three cones to feather the air gradually, to prevent it from contacting the ceiling around the diffuser.
Ensure stability	$1.36 (13%)	Team referred to its function-cost worksheet and found that half of the webs, web clips and rivets, all of the springs, and much of the assembly labor was to ensure stability. Changed to two wireforms with two of the legs spot-welded to a newly designed center cone.
Ease installation and simplify adjustment	$1.07 (10%)	Both of these functions are performed by access areas in the cones, to permit the use of a screwdriver to attach the unit to the ductwork and to adjust the ductwork damper after installation. Installation access areas were deleted, since modern installation does not require screw attachment. A hole was added to permit damper adjustment.
Protect shipments	$0.86 (8%)	The prestudy discussions with users/customers revealed that diffusers were invariably ordered in pairs. The shipping carton was redesigned to carry two units at a significant cost reduction.

team made. In addition, the team responded to installers' complaints by doubling the height of the diffuser collar to simplify installation. This change did not increase cost, since the collar is press-formed from material displaced in forming the hole.

The changes (Figs. 13-37 and 13-38) accomplished the classic value analysis result: The material cost was reduced by 24 percent and the labor cost was reduced by 84 percent saving the company nearly $500,000, while the diffuser performs its task much better in terms of both user needs and wants. Pressure drop is reduced by 20 percent, and air capacity is increased by 30 percent. These improvements, plus a 20 percent price reduction, have resulted in a significant increase in market share.

19. Creation of a Totally New Design
The Product: Plastic strapping machine
Lesson Learned: A total redesign is focused by value analysis

Signode Industries used the value analysis system to create a totally new product. The company has chosen to use the term *value engineering* when the process is used upstream, in the engineering environment; this term will be used in the discussion below.

The product line was a series of strapping machines for packaging, all of which were highly regarded in the industry. The products were succumbing rapidly to price-based European and Far Eastern competition. The company chose to base its value engineering for the proposed Spirit ® machine on its current Model ML machine, a particularly

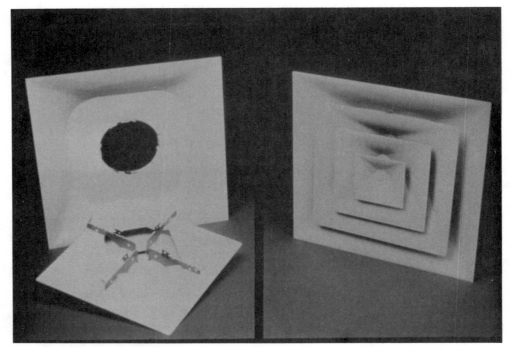

Figure 13–37. The original configuration of the ceiling air diffuser resulted in several failings, from the user's point of view.

rugged and full-featured heavy-duty strapping machine with a prohibitively high manufacturing cost (Fig. 13–39).

A marketing/engineering trade-off effort resulted in a basic machine, stripped of many of the features that marketing felt were essential. Its cost was two-thirds of the ML equipment.

Four multidisciplined teams were set up, including representation from marketing, engineering, purchasing, and manufacturing. They were guided by a set of specifications, supplied by the marketing department, which defined the required features and the cost constraints. They began by creating a FAST diagram. Since the new machine did not yet exist, their basis for definition of the functions was an intimate knowledge of the functions performed by an existing machine that had previously undergone value engineering. Best estimates were made of the cost of the new machine, based on the known costs of the existing one. The teams allocated the costs to the functions. Then they allocated data from a focus panel of machine users and indentified ten value engineering targets. In the creativity and synthesis phases, they focused on those targets, creating a new machine with an estimated product cost that still exceeded their goal.

At this point, a status arbitration session was held. Its participants were the four value engineering team captains, the vice president of marketing, the business unit manager, and the projects manager. This was essentially a repeat of the original trade-off effort, but it benefited greatly from the broadened function-cost viewpoint that the teams developed in their creativity and synthesis phases.

The objective of the session was to discuss and resolve the originally defined value

Figure 13–38. The value-analyzed version of the ceiling air diffuser reduces costs—and price—and performs its task much better.

engineering targets. The marketing representatives clarified the absolute needs that had to be met. They also defined the functional areas in which expectations could be lowered. The team captains presented the new concepts, and a goal cost was jointly declared to be possible for a machine that would be strongly received in the market. A second creativity and synthesis effort then took place based on this broadened base of data.

The result of the effort of these four teams was the totally new Spirit® machine, with a cost that reflected a 39 percent reduction from the estimated cost of the original feature-stripped version (Fig. 13–40). Its features, its reliability, and its ergonomics were superior to anything offered by any competitor in the world, and it quickly went into full and continuing production in the Signode factory in Glenview, Illinois.

Two features distinguish this dramatically successful value engineering study: (1) The objective (and the result) was a totally new machine, with essentially no parts from existing machines; and (2) the status arbitration session, performed during the value engineering study, brought value engineering, engineering, and marketing together in a unique and powerful spirit of cooperation.

It is worth noting that a value analysis study was performed four years later. Benefiting from field experience, it led to results including significant further improvements in product function and reliability, as well as reductions in manufacturing costs.

Figure 13–39. The former standard of the trade, the Signode ML light-duty strapping machine, was dying in the world market; it was overspecified and overpriced. (Courtesy of Signode Corporation.)

20. Product Line Standardization
The Product: Wet horizontal magnetic testing machine
Lesson Learned: Function-cost data leads team to a standardized module

The Magnaflux Corporation performed a value analysis study of model H-810, the flagship of its product line (Fig. 13–41). The objective was to redesign that specific model to satisfy user concerns and reduce its cost.

Two teams accomplished both goals, but also realized a far more fundamental objective with far-reaching benefits, affecting not only the simplicity and cost of manufacturing and distribution of the entire product line, but also the basic manufacturing systems of the organization.

The seeds of the breakthrough had been planted by several management activities in the months before the value analysis study. Efforts had been made to simplify the rather complex product line. A vendor qualification program had been established. There were European pressures to reduce the number of models. A company objective to leapfrog over the competition was defined. All ten members of the two value analysis teams were familiar with these corporate developments.

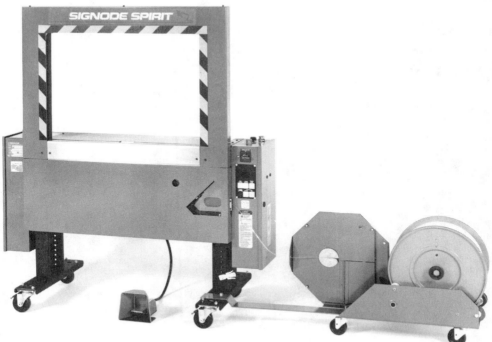

Figure 13–40. The new Signode Spirit® strapping machine, redesigned after value engineering and a status arbitration session, costs much less and has significantly increased the fulfillment of user needs and wants. (Courtesy of Signode Corporation.)

The shift from the initial focus on product cost improvement started during the information phase, in which the teams realized that the anatomy of the function costs of many mechanical assemblies was out of line. Their conviction was catalyzed during the expert/vendor session of the synthesis phase. Most of the mechanical assemblies and weldments are manufactured by outside vendors. Vigorous interchanges with several of them, particularly the major supplier of fabricated assemblies, rapidly led both teams to the realization that a major problem was the lack of standardization in mechanical assemblies. A proposal was made that all mechanical assemblies be simplified.

During the post-synthesis intersession period, the champion of the simplification redesign concept worked with the vendors and the engineering department to develop a new tank/frame weldment, which eliminated major operational and fabrication problems and, much more important, became the standard tank and frame for most of the models in production (Fig. 13–42). This triggered a similar standardization of doors, access panels, and other significant mechanical parts and assemblies. The immediate savings in cost for the model under study were significant (Fig. 13–43), but the major savings were in even greater reductions in the cost of other models.

Additional savings in work-in-process, distribution, stocking, and delivery time were sufficient to affect the Magnaflux bottom line dramatically. The larger volume buys and the predictability of material requirements permitted the company to implement a full just-in-time material control system.

In the final presentation, the teams reported probable savings of 20.2 percent. They

Figure 13–41. The Magnaflux Corporation was the originator and the leader in its field of non-destructive testing; the cost of this flagship model H-810, however, was excessive.

Figure 13–42. The new standardized Magnaflux tank module became a major boon to its manufacturer in unexpected ways.

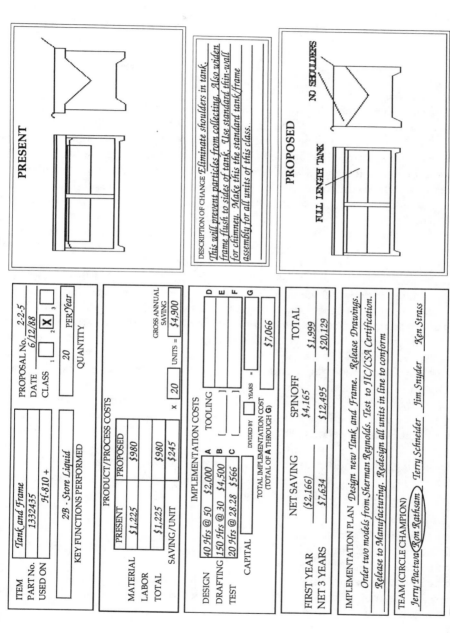

PRESENT

PROPOSED

NO SHOULDERS

FULL LENGTH TANK

ITEM	Tank and Frame	PROPOSAL No.	2-2-5
PART No.	1332435	DATE	6/12/88
USED ON	H-810 +	CLASS	1 ☐ 2 ☒ 3 ☐

KEY FUNCTIONS PERFORMED 2B - Store Liquid

QUANTITY 20 PER Year

PRODUCT/PROCESS COSTS

	PRESENT	PROPOSED
MATERIAL	$1,225	$980
LABOR		
TOTAL	$1,225	$980
SAVING/UNIT		$245

$245 × 20 UNITS = $4,900 GROSS ANNUAL SAVING

DESCRIPTION OF CHANGE *Eliminate shoulders in tank.*
This will prevent particles from collecting. Also widen
frame flush to sides of tank. Use standard thin-wall
for chimney. Make this the standard tank/frame
assembly for all units of this class.

IMPLEMENTATION COSTS

				TOOLING	
DESIGN	40 Hrs @ 50	$2,000	**A**	**D**	
DRAFTING	150 Hrs @ 30	$4,500	**B**	**E**	
TEST	20 Hrs @ 28.28	$566	**C**	**F**	
CAPITAL				**G**	$7,066

DIVIDED BY ☐ YEARS =

TOTAL IMPLEMENTATION COST
(TOTAL OF **A** THROUGH **G**) $7,066

	NET SAVING	SPINOFF	TOTAL
FIRST YEAR	($2,166)	$4,165	$1,999
NET 3 YEARS	$7,634	$12,495	$20,129

IMPLEMENTATION PLAN *Design new Tank and Frame. Release Drawings.*
Order two models from Sherman Reynolds. Test to JIC/CSA Certification.
Release to Manufacturing. Redesign all units in line to conform

TEAM (CIRCLE CHAMPION)
Jerry Pactun ⊙*Ron Rathsam* *Terry Schneider* *Jim Snyder* *Ken Strass*

Figure 13–43. Value analysis proposal 2-2-5 redesigned the basic tank, reducing the cost by $4,900 per year; the new tank also was made a standard for all units in its class, saving an additional $4,165 per year.

supported the presentation with actual models of most of the significant changes. Management receptivity to the proposals is best illustrated by the present position of Ron Rathsam, a team member who was the champion of the effort. He is presently manager of standard products—and a passionate supporter of the value analysis system.

21. Change Chemical Formula
The Product: 718 welding rod, ⅛" diameter
Lesson Learned: Even a welding rod benefits from value analysis

Hobart Brothers Company is a leader in welding systems and supplies. The company elected to value analyze a welding rod, despite some reservations about the efficacy of the value analysis system when applied to an item with only two parts: wire, which is purchased on reels, and a chemical fluxing mixture, which is used to coat the wire.

The value analysis team comprised the true experts on the welding rod: the process engineer, manager of finance, marketing manager, purchasing agent, and development engineer. The FAST diagram (Fig. 13–44), developed by the team in six hours on the first day of an eight-week, 55-hour effort, contained 37 independent functions! Three of these were basic: "conduct electricity," "convert energy," and "deposit metal"; the remainder were supporting.

The team members' meticulous allocation of the function-cost and focus panel attitude data resulted in their identifying 11 areas where they felt that creativity and synthesis effort would pay off. These value analysis targets are shown in Figure 13–45.

In focusing their creativity and synthesis effort on those 11 targeted areas, they developed four major proposals for change. The expected net savings for the first three years totaled $1,400,160. One of these proposals alone, by significantly changing the formula of the coating material, accounted for a $515,525 predicted saving over three years (Fig. 13–46). The formula change was triggered by the combination of 10 of the 11 value analysis targets, and was the result of several weeks of intensive effort by most of the team members, both in their value analysis sessions and during the two three-week intersession periods. The details of this change in formula are, of course, highly proprietary.

At this writing, the action plan is still under way. It is likely, however, that it will be implemented in full, since the team members included all of the key decision makers, including the responsible development engineer.

22. Capital Tooling Improvement
The Product: Manufacturing plant for a new concept of variable-displacement automotive air-conditioning compressor
Lesson Learned: Value analysis succeeds with capital expenditures, too

A division of a major U.S. corporation manufactured automotive air conditioners. It invented a controlled-displacement Freon compressor for the newer, smaller U.S. and foreign automobiles. The unique feature of the unit was the nearly complete absence of complex controls. The displacement of the compressor varies in direct relation to the demand on the air-conditioning system by responding internally to inherent delta pressures. The division's request to the corporate office for permission to spend capital totaling $89,202,000 was initially rejected. The $89 million figure fulfilled all of the corporate "hurdle rates" on ROI and payback, but based on dramatic previous value

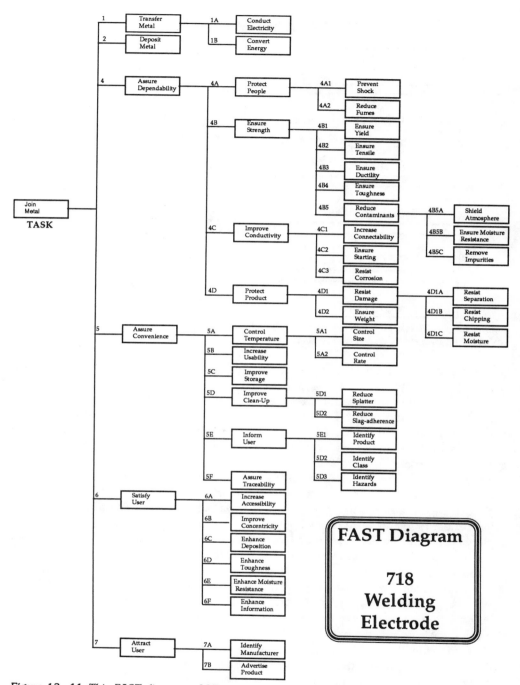

Figure 13–44. This FAST diagram of 37 independent functions represents a simple, flux-coated arc welding rod.

718 Welding Electrode

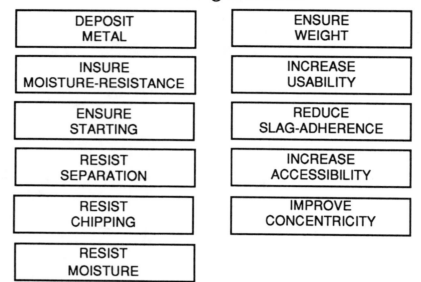

DEPOSIT METAL	ENSURE WEIGHT
INSURE MOISTURE-RESISTANCE	INCREASE USABILITY
ENSURE STARTING	REDUCE SLAG-ADHERENCE
RESIST SEPARATION	INCREASE ACCESSIBILITY
RESIST CHIPPING	IMPROVE CONCENTRICITY
RESIST MOISTURE	

Figure 13–45. *These 11 functions were identified as value analysis targets; the team then came up with four proposals with major expected savings.*

analysis successes, the corporate office felt that the division would greatly benefit by applying value analysis to the design of the manufacturing process.

Three teams were established. Each was led by one of the three men who had team-designed and estimated the $89 million system. The other 12 team members were, without exception, the key decision makers in the division in their areas of specialization. Their goal was to reduce the capital investment by 32 percent and, incidentally, to reduce the number of labor hours to produce a unit by the same percentage. This double-barreled objective required that the team set up two cost structures to allocate. They chose to evaluate not only the capital and the standard hours per unit, but also the tooling and the material per unit. Since material was a relatively small portion of the total cost per unit, and since their experience with value analysis led to a conviction that material cost reduction always results, only the capital, tooling, and standard hours per unit underwent function-cost allocation to the FAST diagram (Table 13–3).

The user data collected in a focus panel was also allocated to the FAST diagram, and the resultant data focused the teams on eight of the functions for their creativity and synthesis phases. Most team members applied well over the requisite 25 percent of their time to investigation and definition of their championed concepts. The results of the value analysis are shown in Table 13–4.

The reduction of over $28 million in capital investment was sufficient to stimulate the corporate office to release the funding quickly. The nearly $3 million reduction in tooling cost was a bonus, as was the reduction in unit cost by $3.69.

A typical value analysis proposal—one of 35 that the three teams completed—is shown in Figure 13–47.

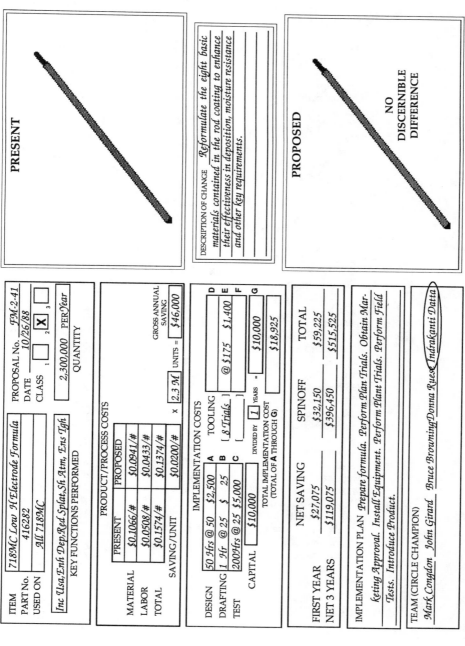

PRESENT

DESCRIPTION OF CHANGE: *Reformulate the eight basic materials contained in the rod coating to enhance their effectiveness in deposition, moisture resistance and other key requirements.*

PROPOSED

NO DISCERNIBLE DIFFERENCE

ITEM: *718MC Low H Electrode Formula*
PART No.: 416282
USED ON: *All 718MC*
KEY FUNCTIONS PERFORMED: *Inc Usa, Enh Dep, Red Splat, Sh Atm, Ens Tgh*

PROPOSAL No. *FM-2-41*
DATE *10/26/88*
CLASS: 1 [ ] 2 [X] 3 [ ]
2,300,000 PER Year QUANTITY

PRODUCT/PROCESS COSTS

	PRESENT	PROPOSED
MATERIAL	$0.1066/#	$0.0941/#
LABOR	$0.0508/#	$0.0433/#
TOTAL	$0.1574/#	$0.1374/#
SAVING/UNIT	$0.0200/#	× 2.3 M UNITS =

GROSS ANNUAL SAVING $46,000

IMPLEMENTATION COSTS

DESIGN	50 Hrs @ 50	$2,500	A
DRAFTING	1 Hr @ 25	$ 25	B
TEST	2000Hrs @ 25	$5,000	C
			D
TOOLING	[8 Trials] @ $175	$1,400	E
			F
CAPITAL	$10,000	$10,000	G

TOTAL IMPLEMENTATION COST (TOTAL OF A THROUGH G) $18,925
DIVIDED BY 1 YEARS

	NET SAVING	SPINOFF	TOTAL
FIRST YEAR	$27,075	$32,150	$59,225
NET 3 YEARS	$119,075	$396,450	$515,525

IMPLEMENTATION PLAN: *Prepare formula. Perform Plan Trials. Obtain Marketing Approval. Install Equipment. Perform Plant Trials. Perform Field Tests. Introduce Product.*

TEAM (CIRCLE CHAMPION)
Mark Congdon John Girard Bruce Browning Donna Ruess (Indrakanti Datta)

Figure 13–46. *Value analysis proposal FM-2-41 modified the coating formula of the welding rod, which saved 2 cents per electrode—for a total annual saving of $46,000.*

TABLE 13–3. FUNCTION-COST STRUCTURE
FOR MANUFACTURE OF NEW COMPRESSOR

Type of Function	Function	Capital	Tooling	Std hrs
Task	Flow refrigerant	$89,391,000	$7,109,000	.5074
Basic	Admit gas	$428,000	$162,000	.0068
	Open port	$392,000	$156,000	.0062
	Lower pressure	$36,000	$6,000	.0006
	Discharge gas	$3,400,000	$180,000	.0122
	Open port	$231,000	$87,000	.0036
	Increase pressure	$3,169,000	$93,000	.0086
	Compress gas	$15,930,000	$1,055,000	.0871
	Close port	$3,421,000	$156,000	.0168
	Reduce volume	$12,509,000	$899,000	.0703
	Transmit torque	$4,872,000	$234,000	.0252
	Convert rotation	$7,637,000	$665,000	.0451
	Restrain motion	$4,304,000	$488,000	.0276
	Follow locus	$3,333,000	$177,000	.0175
	Enclose boundaries	$865,000	$272,000	.0109
	Facilitate mounting	0	0	0
Total basic		$20,683,000	$1,669,000	.1170
Supporting				
	Assure dependability	$41,853,000	$4,252,000	.3052
	Reduce wear	$25,182,000	$2,226,000	.1540
	Reduce friction	$16,165,000	$1,115,000	.0770
	Insure lubrication	$2,220,000	$205,000	.0134
	Cycle unit	$23,000	$9,000	.0028
	Minimize contamination	$6,774,000	$897,000	.0608
	Verify conformance	$1,608,000	$655,000	.0406
	Resist corrosion	$40,000	$23,000	.0013
	Minimize breakage	$5,688,000	$271,000	.0273
	Limit leakage	$8,259,000	$909,000	.0659
	Protect unit	$1,076,000	$168,000	.0161
	Facilitate manufacturing	$3,216,000	$122,000	.0079
	Assure convenience	$5,321,000	$281,000	.0143
	Ease servicability	$2,520,000	$106,000	.0051
	Identify unit	$175,000	$80,000	.0027
	Standardize mount	$2,626,000	$95,000	.0065
	Satisfy user	$18,343,000	$769,000	.0620
	Minimize NVH	$1,112,000	$85,000	.0063
	Maximize efficiency	$3,314,000	$133,000	.0146
	Control displacement	$13,748,000	$478,000	.0361
	Modulate stroke	$12,072,000	$411,000	.0281
	Sense pressure	$1,676,000	$67,000	.0080
	Generate ▲ P	0	0	0
	Limit corrosion	$169,000	$73,000	.0050
	Attract user	$35,000	$16,000	.0010
	Decrease weight	0	0	0
	Reduce size	0	0	0
	Display color	$35,000	$16,000	.0010
Total supporting		$68,768,000	$5,440,000	.3904

TABLE 13–4. FINAL RESULTS OF THE CD COMPRESSOR CAPITAL STUDY

	Capital Investment	Tooling	Standard Hours	Material
Original	$89,202,000	$7,110,000	.5074	
Reduction	$28,751,000	$2,880,000	.1593	$0.2835
			Net unit cost reduction	$3.6918

23. Trailer Drop Stand
The Product: Tongue jack for manufactured housing
Lesson Learned: User data triggers the invention of a new product

A division of a major manufacturer of automotive parts made a jack that was supplied to the mobile home industry. The jack mounts on the towing tongue of the frame used to deliver manufactured houses to their final sites. When the house is moved into place, the jack is hand-cranked to raise the tongue so that the towing vehicle can disconnect and move away. The same product is used in the general mobile home industry for units up to 40 feet in length.

A team of five young, eager, and very qualified employees, including the project engineer, was assigned to value-analyze the jack.

A focus panel was held in Elkhart, Indiana, a center for the manufacture of mobile homes and manufactured housing. Participants included eight members of management and the decision makers from eight of the company's key customers. One of the likes the panel identified was "speed of operation." The panel rated this feature on the standard 1-to-10 scale from two different viewpoints: The first vote rated its importance to the mobile home industry, while the second vote rated its importance to the recreational vehicle industry. The first industry was rated 1 in importance; the second 8; the range of vote was 5 in both cases.

The value analysis team discussed this data during the third-day value analysis target session. It proved to be an idea stimulator for the project engineer. He saw in the data a "fault" being expressed by the RV people. The "fault" was never explicitly described by the users, even during the dislikes portion of the focus panel session—probably, he reasoned, because there were no jacks available from any supplier that solved the "Speed of Operation" problem.

The project engineer felt that the RV people, with their special requirements, were suffering regularly because each time they needed to extend the jack they had to rotate the crank 40 to 50 turns. Since the RV people always use a separate jack to lift the tongue, most of the turns of the crank were merely to "fill the space" between the jack and the ground. He felt that if his company could provide a rapidly deploying jack at a reasonable cost, perhaps it could dominate the market.

Such a redesign, he reasoned, would remove the product from the category of commodity and permit it to compete on value instead of price. This viewpoint fits well with a precept of value analysis, that no product or service should ever be regarded as a commodity. Indeed, Theodore Levitt, now the editor of the *Harvard Business Review,* has written, "there need be no such thing as a commodity" (Peters and Austin 1985, 60).

The project engineer agreed to champion a concept that is described in the value analysis proposal shown in Figure 13–48. The complete redesign solved some strength problems while adding the feature that, by pulling a pin, the jack could drop down to touch the ground. At that point, only a few turns on the crank of the separate jack would lift the tongue.

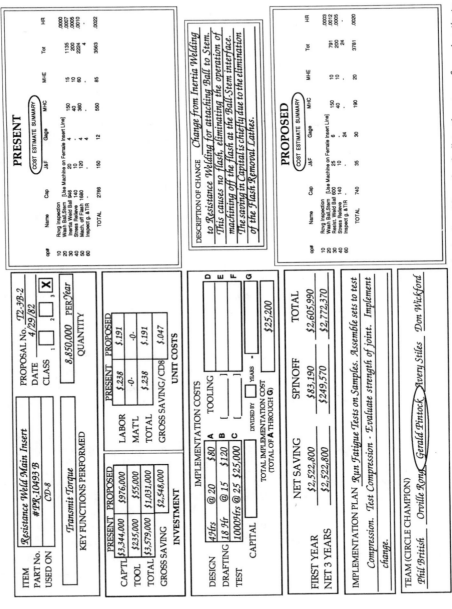

Figure 13–47. *Value analysis proposal T2-3B-2 changes the method of welding the ball to the stem from inertia to resistance welding, saving nearly $2.6 million in capital and reducing the unit cost by 4.7 cents.*

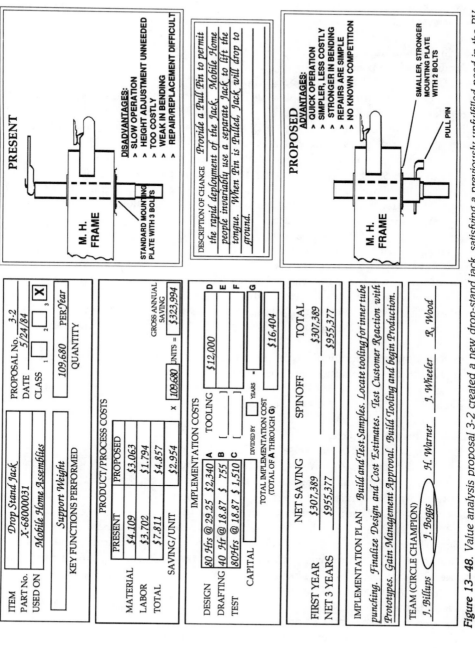

Figure 13–48. Value analysis proposal 3-2 created a new drop-stand jack, satisfying a previously unfulfilled need in the RV industry.

This new jack would be marketed to RV manufacturers only. Its cost is significantly lower than the present unit, permitting a decrease in selling price. Coupled with the absence of competition, the value analysis team predicted a dramatic increase in market share.

24. Minimize Mailings
The Product: Billing procedure
Lesson Learned: Value analysis works on procedures

The Indiana Gas Company, a public utility supplying gas to homes, manufacturers, and retailers throughout central Indiana, established a modern value analysis program. The first thing the company chose to analyze was its billing process, starting with the reading of the meter and ending with the collection of fees.

A powerhouse team of managers and supervisors created a FAST diagram and allocated to it the $9,150,000 annual cost of the billing process. The cost data had been presented to the team in the form of a costed, 28-item block diagram, an extremely abbreviated format. This required the team to devote several hours to an expansion of the costing to reveal its detailed structure. Fortunately, the team was qualified to make accurate estimates of this detail, and the resultant function-cost structure revealed that 11.7 percent of the total cost performed the "render bill" function. Focus panel data revealed a relative insensitivity of the users to the method of billing. The team identified the "render bill" function as a value analysis target based on this data; and the team entered the creativity phase focusing on this function.

The supervisor of employee benefits accepted ownership of a concept for quarterly billing that arose during the team's synthesis phase. Her championed proposal is shown as Figure 13–49.

It is common for value analysis to be applied to improve a procedure, such as the billing process above or design release, laboratory testing, and so on. The function viewpoint and the rigor of the job plan make it the optimum problem-solving system whenever the problem involves costs and functions.

25. Stable Stove-Top Spider
The Product: Restaurant oven/range
Lesson Learned: User data focuses the team on product faults

A manufacturer of commercial kitchen equipment produced a restaurant range that was considered the standard of the trade. The company undertook a value analysis at the suggestion of its corporate parent to ensure future dominance in its market.

The team of six was unusual, in that their work locations were separated by hundreds of miles. Three were from the manufacturing plant on the East Coast; one was from sales in the South, another was from marketing in the Midwest, and the last was from corporate headquarters in the North. The problems of coordination were anticipated and therefore were fairly well handled.

A user/customer focus panel was held in a centrally located city with nine actual users and six key marketing, sales, and engineering people from the producing company. The panel identified 35 features or faults, rating each on the 1-to-10 scale. One of the likes and two of the dislikes are shown in Figure 13–50.

Feature 14 was rated as very important. Clearly the users feel that easy removability of the cast spider grates is a feature that must not be degraded. Faults 10 and 11, however, reflect a pair of serious complaints. During the discussion before they voted

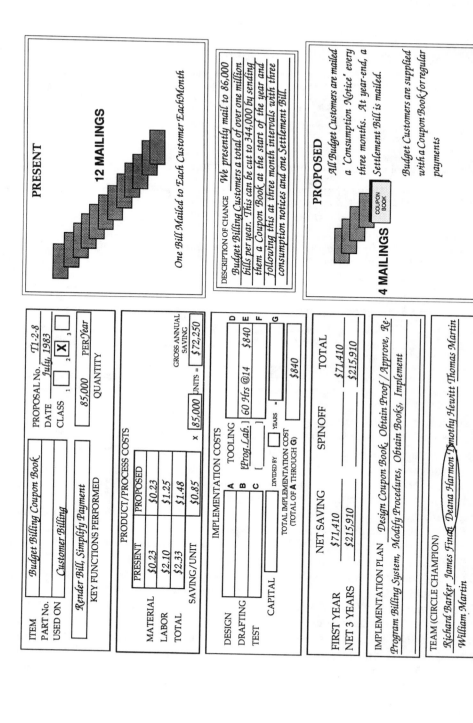

PRESENT

12 MAILINGS

One Bill Mailed to Each Customer Each Month

DESCRIPTION OF CHANGE *We presently mail to 86,000 Budget Billing Customers a total of over one million bills per year. This can be cut to 344,000 by sending them a Coupon Book at the start of the year and following this at three month intervals with three consumption notices and one Settlement Bill.*

PROPOSED

All Budget Customers are mailed a 'Consumption Notice' every three months. At year-end, a Settlement Bill is mailed.

COUPON BOOK

Budget Customers are supplied with a Coupon Book for regular payments

4 MAILINGS

ITEM	*Budget Billing Coupon Book*
PART No.	
USED ON	*Customer Billing*

PROPOSAL No. *T1-2-8*
DATE *July 1983*
CLASS 1 ☐ 2 ☒ 3 ☐

Render Bill, Simplify Payment
KEY FUNCTIONS PERFORMED

85,000   PER *Year*
QUANTITY

PRODUCT/PROCESS COSTS

	PRESENT	PROPOSED
MATERIAL	$0.23	$0.23
LABOR	$2.10	$1.25
TOTAL	$2.33	$1.48
SAVING/UNIT		$0.85

× 85,000 UNITS = $72,250

GROSS ANNUAL SAVING $72,250

IMPLEMENTATION COSTS

A		**TOOLING** D
DESIGN		
B	*Prog. Lab.* 60 Hrs @14 $840 E
DRAFTING		
C	[ ] F
TEST		
CAPITAL		

DIVIDED BY ☐ YEARS = $840 G

TOTAL IMPLEMENTATION COST (TOTAL OF **A** THROUGH **G**)

	NET SAVING	SPINOFF	TOTAL
FIRST YEAR	$71,410		$71,410
NET 3 YEARS	$215,910		$215,910

IMPLEMENTATION PLAN *Design Coupon Book, Obtain Proof / Approve, Re-program Billing System, Modify Procedures, Obtain Books, Implement*

TEAM (CIRCLE CHAMPION)
Richard Barker James Finch Deana Harmon Timothy Hewitt Thomas Martin William Martin

Figure 13–49. *At a public utility, value analysis proposal T1-2-8 changed the payment schedule, for a gross saving of $72,250 per year.*

LIKES (Features or Characteristics)

	1	2	3	4	5	6	7	8	9	10	11	12	13	14	15	MODE
14 Spider Grates are easily removable	7	9	6	7	10	5	-	9	9	9	6	9	9	8	-	9R5

DISLIKES (Faults or Complaints)

10 The Spider Grates move and tilt when trying to move pots around the top of the Range.	10	10	10	8	8	8		10	9	10	8	9	9	10	9	8	10R2
11. Spider Grates easily break	2	10	10	-	9	6		8	6	3	9	9	-	10	10	5	10R8

[RESPONDENTS 1 THROUGH 6 ARE MANAGERS.
RESPONDENTS 7 THROUGH 15
ARE USER/CUSTOMERS]

Figure 13–50. An excerpt from focus panel data about the restaurant range shows one like and two dislikes related to the spider grates.

ITEM	FUNCTIONS					
	3	4D	6C	6D	7A	7B
	Support Weight	Increase Strength	Enhance Efficiency	Enhance Flexibility	Improve Appearance	Exude Strength
Spider Grate $19.92	$6.00	$10.92	-0-	-0-	$3.00	-0-

Figure 13–51. Six of the functions on the FAST diagram related to the spider grate area; each of these function costs is tabulated on this worksheet excerpt.

on the seriousness of the fault, several users described the difficulty of moving a pot from one burner to another. During such a move, whenever the pot is positioned on the edge of a spider, the grate becomes unstable, occasionally causing the pot to tip over. In addition, the radiating "fingers" of the spider grates broke off easily when the grate was dropped on the floor. The company was doing a profitable business in replacement spiders, but the users were unhappy.

During their meticulous analysis of function cost, the team members allocated the $19.92 cost of the spider grates to three different functions, as shown in Figure 13–51. The logic used in this allocation of the spider grate's cost among six related functions on the FAST diagram followed the reasoning shown in Table 13–5.

In the value analysis target session, the team members took note of the significant faults of the spider grate. Then they added together each of the total FAST diagram costs for the grate-related functions as shown in Figure 13–52. The total was $119.00, over 30 percent of the total cost of the range. On these combined bases, they declared the "support weight" function a value analysis target. The creativity and synthesis phases resulted in 85 concepts, one of which led to the value analysis proposal shown in Figure 13–53. This proposal was championed by the team member from corporate headquarters, who has since tooled and implemented the change.

TABLE 13–5. LOGIC APPLIED BY TEAM TO ALLOCATE SPIDER GATE COST

Function and Cost	The team's reasoning
Increase strength ($10.92)	The spider grates are cast with a thicker web than necessary to just barely hold weight. They were designed to resist all normal shock loads during use.
Enhance efficiency (0)	The spider grate requires more finishing labor because of the complexity of its configuration. This enables more of the flame to contact the pot, enhancing efficiency.
Enhance flexibility (0)	The very fact that the spider grates can be removed makes them more flexible in application. There is, however, no added cost in the grate to accomplish this function.
Improve appearance ($3.00)	The surface of the grate is polished. This contributes nothing to performance, only to appearance.
Exude strength (0)	In addition to being strong, the grates perform the added function of appearing strong. No cost is added to the grate to accomplish this function, so its function-cost is listed as zero.
Support weight ($6.00)	The prime reason for the existence of the spider grates is to support the weight of the pots on top of the range. The remaining $6.00 is therefore allocated to this function.

3	Support Weight	$27
4D	Increase Strength	$46
6C	Enhance Efficiency	$ 5
6D	Enhance Flexibility	$ 6
7A	Improve Appearance	$28
7B	Exude Strength	$ 7
	Total	$119

Figure 13–52. Total FAST diagram costs for the functions listed in Figure 13–51.

26. Foam Pattern Casting
The Product: Precision multiple hydraulic valves
Lesson Learned: Vendor input pays off

A hydraulic valve manufacturer applied value analysis to its entire 10000 Series line of stacked multiple controls. The chief application for its valves is in mobile earth-moving equipment, a market that had recently weakened. The three value analysis teams were given a goal of reducing the cost of the four-unit valve by 10 percent after removing all user-perceived faults. At the kickoff session, the president presented these goals to the teams and rationalized the goals by describing the threats being mounted by competition, particularly with the company's key customer, a major tractor manufacturer.

The teams accomplished their goal, presenting 34 detailed change proposals that reduced the four-unit cost by 11 percent, for a total cost saving of $936,000, including $440,000 in spin-off savings. As in the usual value analysis effort, many of the proposals were prosaic, a few were inventive, and one of two were major breakthroughs. This case study discusses one of the more prosaic proposals, saving only $1.99 per valve. It is significant, however, because the concept on which it was based did not emerge in

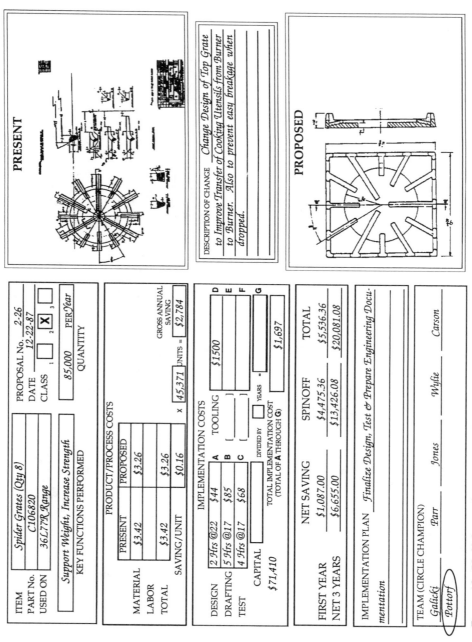

Figure 13–53. Value analysis proposal 2-26 totally redesigned the spider grates, correcting the user-perceived faults and, as a bonus, reducing the grates' cost by $2,784 per year.

the function-oriented information phase, nor in the function-directed creativity phase that followed.

The team entered the vendor session portion of the synthesis phase, at noon on the fifth day of value analysis sessions, with 88 championed concepts. A group of experts and vendors had been invited to spend four hours with the teams serving as sounding boards for the teams' concepts. The rules for inviting vendors had included the caveat: "Be sure that one-half of them are not present vendors. Invite some hungry outsiders." One such vendor was from a foundry specializing in foam pattern molding. This was a new process to the team members, and one of them saw great potential in changing the valve bodies from sand casting, with its many secondary machining operations, to foam pattern casting, with its near-net-shape capability and its ability to cast undercuts. He volunteered to champion the application of the new process to the 10000 Series valve.

One of the proposals resulting from this investigation is shown in Figure 13–54. The company's success with this change established foam pattern as the standard process for all of its complex castings.

27. Drive-Through Rear Axle
The Product: High-travel forklift
Lesson Learned: User data and function definition lead to a creative solution

The forklift was a key product of the construction equipment division of a major international corporation. The high-travel forklift is used to move materials from the ground to a higher floor of a building under construction or renovation. A corporate-funded program of value analysis was established and, since its highly accepted forklift had a cost problem, it was chosen as the first product to undergo the process.

The three teams that were established each created a FAST diagram of the complete high-travel forklift. In a melding session, all three teams joined to develop a single FAST diagram with the combined viewpoints of all participants. Then they costed the diagram meticulously and posted the data from a user/customer focus panel.

In the analysis phase, they identified function 5A3, "increase strength," as a value analysis target. Its $2,771 cost was 17 percent of the total cost of the forklift, and two serious faults were allocated to "increase strength": One of the faults related to failure of the boom, but the other, with a seriousness of 8, was expressed by the focus panel as "axle housing cracks." This led the teams, in their creativity and synthesis phases, to conceive a new drive arrangement that eliminated the costly and relatively fragile long drop transmission.

The new concept was presented to an outside company engineer during the expert/vendor interchange. It developed that the new transmission and axles required were available. The chief engineer championed the redesign (Fig. 13–55).

This rather dramatically different driveline concept evolved only through the rigorous procedure constrained by the job plan. The teams first shifted their viewpoint to function. Then they costed the functions, revealing a 17 percent concentration in the "increase strength" area. Their user data revealed a fault in that same area, specifically on the transmission, focusing their problem solving on the driveline. The unconstrained creativity phase and the search orientation of the synthesis phase, strengthened by the subsequent verification of the engineer from the transmission supplier, resulted in a three-year net saving of over one-third of a million dollars.

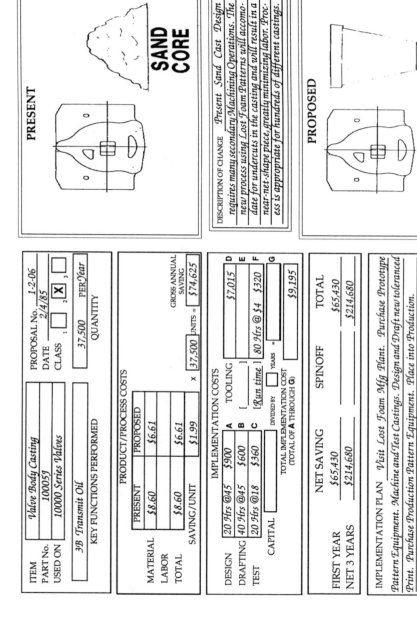

Figure 13–54. *Value analysis proposal 1-2-06 changed the company over to foam pattern casting, for a gross saving of $74,625 per year.*

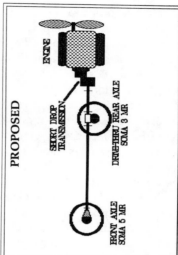

PRESENT

ENGINE

LONG DROP TRANSMISSION

REAR AXLE

FRONT AXLE

DESCRIPTION OF CHANGE *Replace present relatively fragile Long Drop Transmission with a Short Drop Transmission from Soma. Also replace both axles. The Front Axle will be a Soma 5MR. The Rear Axle will be a Drive-Thru Soma 5MR.*

PROPOSED

ENGINE

SHORT DROP TRANSMISSION

DRIVE-THRU REAR AXLE SOMA 3 MR

FRONT AXLE SOMA 5 MR

ITEM	*Long Drop Driveline*	PROPOSAL No. *2-3-1*
PART No.	*Various*	DATE *10/19/84*
USED ON	*Model 9038*	CLASS 1 [ ] 2 [ ] 3 [X]

KEY FUNCTIONS PERFORMED *Transfer Power, Direct Power*

QUANTITY *150* PER *Year*

PRODUCT/PROCESS COSTS

	PRESENT	PROPOSED
MATERIAL	$3,805.00	$2,863.00
LABOR	$10.00	
TOTAL	$3,815.00	$2,863.00
SAVING/UNIT	$952.00	

$952.00 × 150 UNITS = $142,800 GROSS ANNUAL SAVING $142,800

IMPLEMENTATION COSTS

DESIGN	640 Hrs @$25 $16,000	**A**	TOOLING	$5,000 **D**
DRAFTING	400 Hrs @$15 $ 6,000	**B**	[Material]	$30,000 **E**
TEST	1,000 Hrs @$15 $15,000	**C**	[Labor] 773 Hrs @$11 $8,500	**F**
CAPITAL	[ ]			

DIVIDED BY [ ] YEARS = [ ] YEARS

TOTAL IMPLEMENTATION COST (TOTAL OF **A** THROUGH **G**) $80,500 **G**

	NET SAVING	SPINOFF	TOTAL
FIRST YEAR	$ 62,300		$ 62,300
NET 3 YEARS	$347,900		$347,900

IMPLEMENTATION PLAN *Design complete new Lower Driveline System. Build two complete models. Test and Evaluate. Approve and Release.*

TEAM (CIRCLE CHAMPION)
(Clyde Maki) Allan Doers Paul Poirier Mel Burkholz Larry Demchuk

Figure 13–55. Value analysis proposal 2-3-1 changes the high-travel fork lift from a long-drop to a short-drop transmission. The gross annual saving is $142,800.

237

28. Project Organization Chart
The Product: Butler Construction Company custom order flow
Lesson Learned: Improved project communication improves value

Butler Manufacturing Company established and vigorously supports a company-wide value analysis system. Several value analysis workshops are held each year, with three to four different divisions providing teams and products. Initially, the system tackled hardware products with dramatic success. More recent studies have been on procedures such as the custom order flow or primary paint process. The Butler Construction Company, a division that designs and constructs buildings, elected to value-analyze order flow and project management.

The value analysis team comprised a project manager, a senior construction superintendent, the project engineering manager/buildings, the manager of subcontracting, and a senior district sales manager.

Their costed FAST diagram revealed two significant function costs, one appearing low, the other high (Fig. 13–56).

There had been little specific user/customer reaction in a focus panel held earlier, so the team was faced with the task of identifying value analysis targets from function costs alone. The analysis phase discussion resulted in both "interface resources" and "direct communication" being identified as targets, since the team saw the two as being intertwined.

In the creativity phase and the subsequent synthesis phase, this interrelationship of the two concepts developed into a review of the authority structure of a construction project. The project engineering manager/buildings volunteered to champion a project to define the present authority structure and propose a new one that would ensure resource interfacing through redirecting communications. The team's proposal is shown in Figure 13–57.

The change was implemented immediately. It was one of 22 that the team made, most of which eliminated user-perceived weaknesses in the way butler performed its construction function. Each of the proposals also saved money, with first-year savings for all 22 changes totaling $1,673,000.

29. Bolted Hanger to Header Connection
The Product: Post office conveyor erection process
Lesson Learned: One-shot construction project benefits from modern value analysis

The Mid-West Conveyor Corporation is a division of that giant in value analysis, Philips Industries Inc. The division's product is material handling systems. Its first venture into modern value analysis involved a contract for erecting a conveyor system in a U.S. post office in Dallas, Texas. The total labor content of the contract was nearly $1 million. The value analysis team was charged with accomplishing these four precise goals:

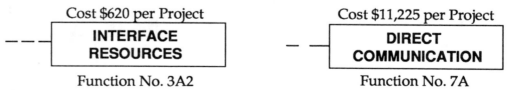

Cost $620 per Project

INTERFACE RESOURCES

Function No. 3A2

Cost $11,225 per Project

DIRECT COMMUNICATION

Function No. 7A

Figure 13–56. A portion of the costed FAST diagram shows two significant function costs, which were determined to be interrelated.

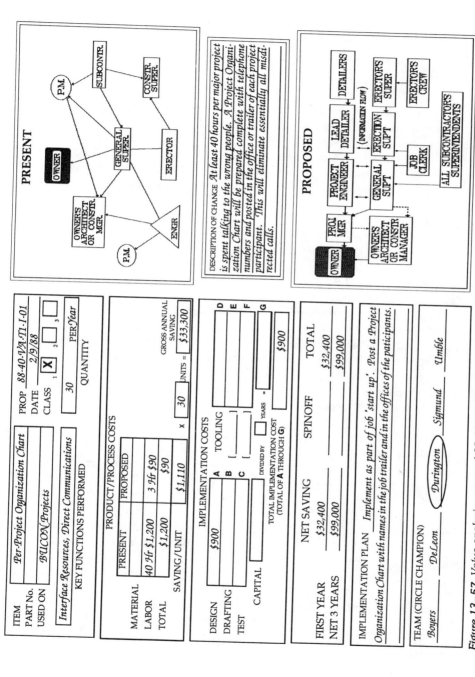

Figure 13–57. Value analysis proposal 88-40-VA-TI-I-01 restructured the communication links in a major project, for a gross saving of $33,300 per year.

239

1. Reduce the erection time by 6,000 hours, equal to a cost saving of $135,000.

2. Save overhead cost in the amount of $14,000.

3. Identify 30 areas for improvement of installation productivity.

4. Identify five areas to take advantage of moving work into the shop, reducing cost by $15,000.

The team exceeded the first two goals and fell only slightly short on the others. The total dollar saving was $287,089, but the team's outstanding accomplishment was in the simplification of the erection effort itself. This is just one of the 27 proposals that the team presented to management. Many involved the modification of procedures or documentation and contract changes with the Postal Service, but the majority, such as the proposal shown in Figure 13–58, were changes in design that simplified the erection process.

The standard value analysis proposal form was modified to accommodate the one-shot nature of the project. All costs are totaled per project. It is likely that additional contracts for similar erections will be performed, further multiplying the savings.

30. Set Temporary SPC Limits
The Product: Overhead and SG&A Expenses
Lesson Learned: Function analysis of chart of accounts identifies value analysis targets

At a major division of an international automobile parts manufacturer, the chart of accounts listed nearly $12 million in annual expenses that were identified as overhead (O/H) or selling, general and administration (SG&A).

The division had experienced a dramatically successful value analysis effort several years before but now was faced with the infamous mid-1980s economic and structural pressures that damaged or destroyed so many Detroit-type businesses. It was also facing a major three-year union contract discussion, which would involve rationalizing to its employees a number of wage and benefit policy changes that had been under consideration. The division established two value analysis teams to attack O/H and SG&A. One team comprised five key employees from administration; this was called the front office team. The other team comprised five key employees from the factory; it was called the plant team.

A focus panel was set up to establish likes and dislikes. The users/customers in this case were all internal to the division: the manager of quality assurance; cost estimator; supervisor; facilities supervisor; inventory control supervisor; buyer; manager of engineering services; accounting; plant engineer; and the manufacturing engineering supervisor.

These people are the internal users of each of the functions funded by the line items of the chart of accounts designated as O/H or SG&A. Their likes and dislikes are the most direct measure of the worth of these functions. Three of the faults or dislikes expressed and rated by this panel are shown in Figure 13–59.

The teams allocated these three faults to different functions on the FAST diagram, so they were not the primary data that the teams used in identifying their value analysis targets; their significance came later. That primary data was the function-cost structure of the FAST diagram.

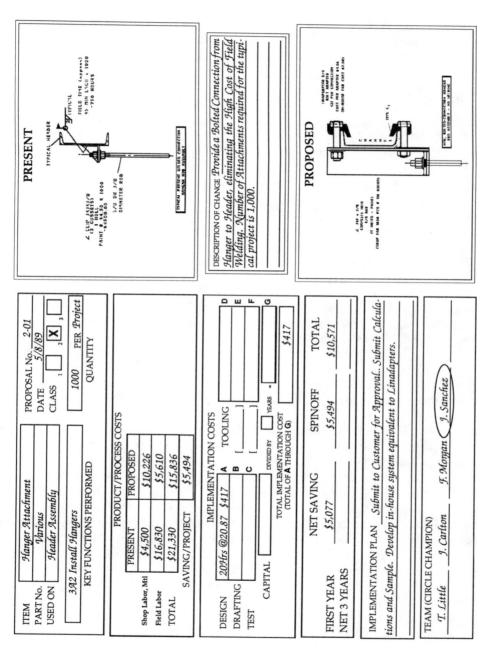

PRESENT

TYPICAL HEADER

⌐TYPICAL

FIELD TIME (approx.)
4> MIN L'XCH × 1000
≈750 HOURS

5/U DK S/U
DIAMETER ROD

4 LVP ASSEN S/U
(2 WISHERS)
1 ROLL
PAINT @ $4.50 × 1000
≈$4500.00

TYPICAL PRESENT HANGER CONNECTION
DESIGN FOR ASSEMBLY

DESCRIPTION OF CHANGE *Provide a Bolted Connection from Hanger to Header, eliminating the High Cost of Field Welding. Number of Attachments required for the typical project is 1,000.*

PROPOSED

|ADAPTER 3/4
BOLT ADAPTER
G2) PER CONNECTION
COSTS PER ADAPTER $1.56
ON-BOUND PER COST $7.04)

4 FW × 1/W
COMPLETE MTN
1/2 BAR
(Y WASHE - PAINT)
CROP PAR 1056 PCS @ 156 BOARD

THIS. MIG TES-PERFECTOR HANGER
DES ACCESSORY - NO WELDING

ITEM	*Hanger Attachment*		PROPOSAL No. *2-01*
PART No.	*Various*		DATE *5/8/89*
USED ON	*Header Assembly*		CLASS [1] [X₂] [3]
			1000 PER *Project*
			QUANTITY

3A2 Install Hangers
KEY FUNCTIONS PERFORMED

PRODUCT/PROCESS COSTS

	PRESENT	PROPOSED
Shop Labor, Mtl	$4,500	$10,226
Field Labor	$16,830	$5,610
TOTAL	$21,330	$15,836
SAVING/PROJECT		$5,494

IMPLEMENTATION COSTS

DESIGN	A 209 hrs @20.87 $417	TOOLING	D
DRAFTING	B [ ]		E
TEST	C [ ]		F
	DIVIDED BY [ ] YEARS =		
CAPITAL			G $417

TOTAL IMPLEMENTATION COST G
(TOTAL OF **A** THROUGH **G**)

	NET SAVING	SPINOFF	TOTAL
FIRST YEAR	$5,077	$5,494	$10,571
NET 3 YEARS			

IMPLEMENTATION PLAN *Submit to Customer for Approval. Submit Calculations and Sample. Develop in-house system equivalent to Linadapters.*

TEAM (CIRCLE CHAMPION)
T. Little   J. Carlton   F. Morgan   J. Sanchez

Figure 13–58. *Value analysis proposal 2-01 was for a totally new hangar attachment process for hanging the conveyor elements from the headers.*

FAULT or COMPLAINT	SERIOUSNESS										MODE
	1	2	3	4	5	6	7	8	9	10	
4 Lack of Plant Utilization	10	10	10	10	10	10	10	8	9	10	10R2
8 Don't do Preventive Maintenance	8	9	-	8	9	10	9	10	9	10	9R2
18 Too many defects in the product	10	9	8	10	10	10	10	10	-	-	10R2

Figure 13–59. Three related faults expressed by the user/customer focus panel led the value analysis team to look critically at waiting time costs.

The team had allocated to the FAST functions each microelement of each line item of cost from the previous year's actual O/H and SG&A expenditures. This structure revealed that eight of the functions each exceeded 4 percent of the total cost. One of these was "insure accuracy," at $1,592,000, or 5.4 percent of the total. In the creativity and synthesis phases, the plant team focused on this function. The manufacturing engineering supervisor suggested that perhaps changes were appropriate in his statistical process control system to improve the value of this function. He saw the three faults identified in Figure 13–58 as being related to the function "insure accuracy." This reasoning led him to champion the proposal shown in Figure 13–60.

The change was implemented immediately. Labor savings for the first year actually exceeded the $65,000 estimate. The structure of the SPC program was rewritten to incorporate this obvious but elusive concept.

31. Eliminate Staples and T Nails
The Product: Insl-Wall® system by the Lester Division of the Butler Manufacturing Company
Lesson Learned: Cost reduction improves appearance

At a corporate-sponsored value analysis workshop, a team from the Lester division analyzed its proprietary Insl-Wall® system.

The FAST diagram contained 38 independent functions, only three of which were classified as basic. A meticulous, one-day allocation of the cost of a typical Insl-Wall® installation revealed that the basic functions were performed by only 5.6 percent of the total cost. Another 58 percent of the cost was concentrated in the "assure dependability" function grouping, while only 10 percent contributed to the grouping "please senses." This would tend to direct the team's problem solving toward the 21 functions under "assure dependability," and indeed the team accomplished a dramatic value improvement with its 25 value analysis proposals, the majority of them focused on dependability targets.

The team also found, however, that an opportunity existed in the attractiveness area. Focus panel data showed that the voters generally rated as a significant dislike the high number of exposed fasteners on the bottom strips on the interior of the panel joints. This was the trigger that directed the team toward a minor but significant change in fastening technique—especially when combined with data from the costed FAST diagram indicating that only 8.4 percent of the total Insl-Wall® cost was devoted to the "enhance appearance" function. Here was an obvious value analysis target: low func-

PRESENT

SPC requires that the operator maintain machine capability within demonstrated limits. . . .

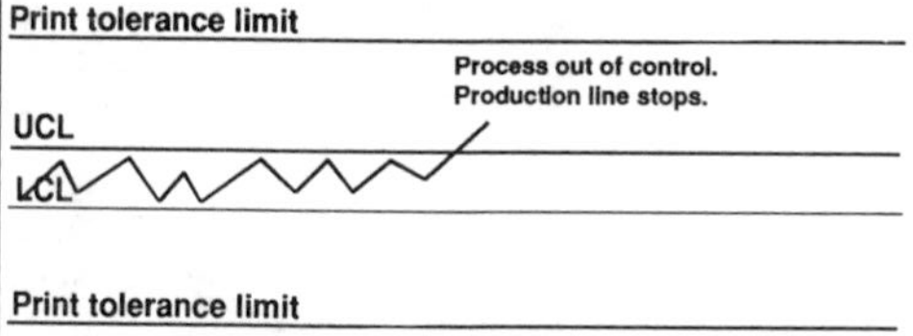

Print tolerance limit

Process out of control.
Production line stops.

UCL
LCL

Print tolerance limit

. . . and if the Equipment exceeds this limit, the Line or the Equipment stops for correction. This often results in a shutdown until the second shift.

DESCRIPTION OF CHANGE *Present SPC procedure: When a process shifts beyond the Lower Control Limit (LCL) or the Upper Control Limit (UCL), the machine is shut down, machine repair notified, and the operator goes on waiting time. Establish temporary limits at which point machine repair schedules second shift repair.*

PROPOSED

When a process exceeds the Temporary Limits, SPC approves continuing until it is convenient to service without stopping line.

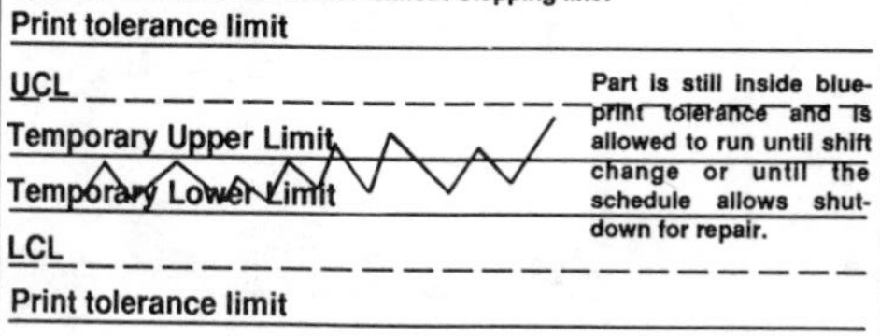

Print tolerance limit

Part is still inside blue-print tolerance and is allowed to run until shift change or until the schedule allows shut-down for repair.

UCL
Temporary Upper Limit
Temporary Lower Limit
LCL

Print tolerance limit

ITEM	SPC Standards Control Restructure	PROP	102
PART No.	All Production	DATE	2/28/86
USED ON		CLASS	[X]1 [ ]2 [ ]3

KEY FUNCTIONS PERFORMED

Insure Accuracy

PRODUCT/PROCESS COSTS

	PRESENT	PROPOSED
MATERIAL		
LABOR	$180,000	$115,000
TOTAL	$180,000	$115,000
GROSS ANNUAL SAVING		$65,000

Present "Waiting Time" charges per year are $180,000. This Proposal could reduce this waste by $65,000

IMPLEMENTATION COSTS

A TOOLING

DESIGN		D
DRAFTING	B [ ]	E
TEST	C [ ]	F
CAPITAL	DIVIDED BY [ ] YEARS =	G

TOTAL IMPLEMENTATION COST
(TOTAL OF **A** THROUGH **G**)

	NET SAVING	SPINOFF	TOTAL
FIRST YEAR	$65,000		$65,000
NET 3 YEARS	$195,000		$195,000

IMPLEMENTATION PLAN *Implementation will be immediate upon approval by Management.*

TEAM (CIRCLE CHAMPION)
J. Eldridge   R. Nigg   R. Phillips   B. Schmidt   (J. Svendor)

Figure 13–60. *Value analysis proposal 102 places an extra set of process limits on each machine operation, triggering a machine repair order before the process exceeds its print tolerance limits; this change saves $65,000 per year.*

tion cost and high function need. After considerable team discussion, a champion prepared and presented the value analysis proposal shown in Figure 13–61.

The average cost of the old fasteners was $4.63 per thousand, while the cost of the brad is $3.55 per thousand. This significant net saving of nearly $5,000 in the first year represents a simple ROI of 561 percent and a break-even period of only slightly over two months. The major reason for the immediate implementation of this change, however, was the improvement in the appearance of the completed panels.

32. Shiplap Panel Joints
The Product: Insl-Wall® System by the Lester Division of the Butler Manufacturing Company
Lesson Learned: User need to "facilitate erection" of the panels results in cost reduction and improved acceptance.

The same team that accomplished the change in brads for panel mounting as described in case study 31 was directed toward function 5B3, "facilitate erection," by the three user responses in the focus panel data shown in Figure 13–62.

The focus panel rated these three characteristics very important, with the mode of their votes being, 8, 10, and 8, respectively. The value analysis team allocated the first two characteristics to function 5B3, "facilitate erection." The third was allocated to the "enhancement" function 6A, "increase flexibility," which the team felt was directly related to function 5B3.

In the analysis phase, the team members noted the focus panel's strong emphasis on the need to facilitate erection. They also noted that the cost of that function was only $34.20, or 8.4 percent of the total system cost. They defined "facilitate erection" as a value analysis target and devoted a 25-minute brainstorming session to the recording of 185 words and phrases on their flip chart. One of the phrases was "eliminate cleanup." This was elevated to "change tongue-and-groove to shiplap."

The team member who made that suggestion prepared the value analysis proposal shown in Figure 13–63.

The resulting $25,000 annual saving is, in a sense, a secondary benefit. The company expects major returns from the increased user acceptance resulting from this change and the majority of the other value analysis proposals reported to management at the final presentation.

33. Reinforcement Sleeve
The Product: Butler Buildings Division Delta Joist®
Lesson Learned: Broadened viewpoint of function-cost and function-worth leads way to product improvement.

Butler had created "the world's finest roof system" for use on hardwall, single-story buildings. Demand for this dramatic new product had far exceeded its expectations. The corporation called for a halt in a major plan for expansion of manufacturing facilities for its new Delta Joist® system in order to permit a value analysis of the system.

Five decision makers from the three Delta Joist® manufacturing plants were formed into a value analysis team. The team created a FAST diagram consisting of 29 independent functions. Each element of the $54,000 cost of a 78,000-square-foot building was allocated to its applicable functions. A focus panel was established comprising builders, architects, engineers, erectors, superintendents, and key decision makers from the

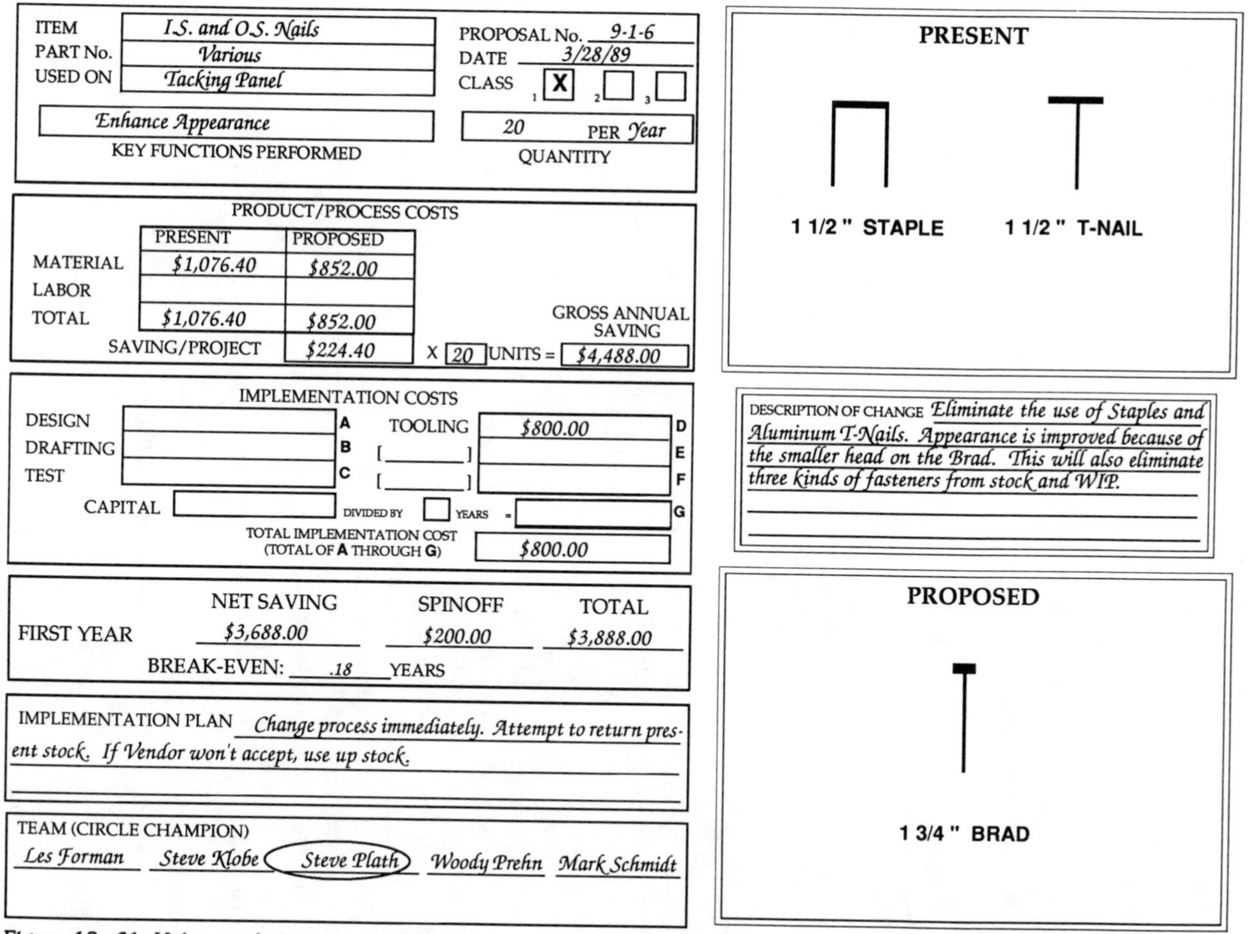

PRESENT

1 1/2 " STAPLE   1 1/2 " T-NAIL

DESCRIPTION OF CHANGE *Eliminate the use of Staples and Aluminum T-Nails. Appearance is improved because of the smaller head on the Brad. This will also eliminate three kinds of fasteners from stock and WIP.*

PROPOSED

1 3/4 " BRAD

ITEM	*I.S. and O.S. Nails*		PROPOSAL No.	*9-1-6*
PART No.	*Various*		DATE	*3/28/89*
USED ON	*Tacking Panel*		CLASS 1 [X] 2 [ ] 3 [ ]	

Enhance Appearance | *20* PER *Year*
KEY FUNCTIONS PERFORMED | QUANTITY

PRODUCT/PROCESS COSTS

	PRESENT	PROPOSED
MATERIAL	$1,076.40	$852.00
LABOR		
TOTAL	$1,076.40	$852.00
SAVING/PROJECT	$224.40	

x [20] UNITS = [$4,488.00]   GROSS ANNUAL SAVING

IMPLEMENTATION COSTS

A	DESIGN		D	TOOLING	$800.00
B	DRAFTING		E		
C	TEST		F		
			G		$800.00

DIVIDED BY [ ] YEARS =

TOTAL IMPLEMENTATION COST (G)
(TOTAL OF A THROUGH G)

	NET SAVING	SPINOFF	TOTAL
FIRST YEAR	$3,688.00	$200.00	$3,888.00

CAPITAL

BREAK-EVEN: *.18* YEARS

IMPLEMENTATION PLAN *Change process immediately. Attempt to return present stock. If Vendor won't accept, use up stock.*

TEAM (CIRCLE CHAMPION)
Les Forman   Steve Kobe (Steve Plath)   Woody Prehn   Mark Schmidt

Figure 13–61. *Value analysis proposal 9-1-6 replaced staples and T-nails with long brads, improving panel appearance and saving $4,488 per year.*

245

	LIKES (Features or Characteristics)																
	1	2	3	4	5	6	7	8	9	10	11	12	13	14	15	16	MODE
5 The openings for Ventilation are Framed Openings, done at Factory.	8	8	9	5	2	1	8	1	10	5	6	9	3	10	3	-	8R9
6 The Panel Design allows for less on-site labor, resulting in fast and efficient erection.	10	9	9	10	8	8	10	10	10	10	9	10	8	10	10	-	10R2
17 The Panels are easily modified; additional openings can be made easily.	8	4	6	5	5	1	6	8	6	6	8	10	6	8	9	9	8R9

Figure 13–62. An excerpt from focus panel data on the Insl-Wall shows three likes that focused the attention of the value analysis team on the "facilitate erection" function.

building division. In an intensive six-hour session, these "users" defined and rated 34 likes and dislikes.

The team converted this data into function-acceptance indices, which it then cross-correlated with function costs for each of the 29 functions. The result was the identification of eight value analysis targets, or areas of opportunity. The team concentrated its creativity and synthesis efforts on these eight functions and proposed 11 changes, which improved user acceptance and reduced the cost by $504,000.

One of the changes was triggered by an unusually high function cost, that of "increase strength," which was $7,600, or 14 percent of the total. The team reviewed the user focus panel data and found that not one of the 34 focus panel responses related to that function. This would seem to imply that either the users don't care about strength or strength is important but such an obvious component of the product that the panel just did not bring it up. The team decided that the latter applied, and resolved to focus its problem solving on modifying the design to increase strength at lower cost while meticulously maintaining existing user acceptance.

This led the team to the different methods used by each plant to reinforce the structural rods. Figure 13–64 is one of the value analysis proposals that resulted from this approach.

The motivation of the team was to improve the product for the user/customer. The result was a significant cost reduction while maintaining the image and the reality of strength. A bonus was the improvement of the appearance of the rod for architectural applications. For all of these reasons, implementation was immediate.

34. Concrete Pouring Shield Adjustment
The Product: Trenchduct® underfloor electrical distribution
Lesson Learned: User viewpoint guides the team to a combined cost reduction and product improvement.

The Walker Division of Butler Manufacturing sent a team of five key employees to a series of corporate value analysis sessions. The project was a major division product

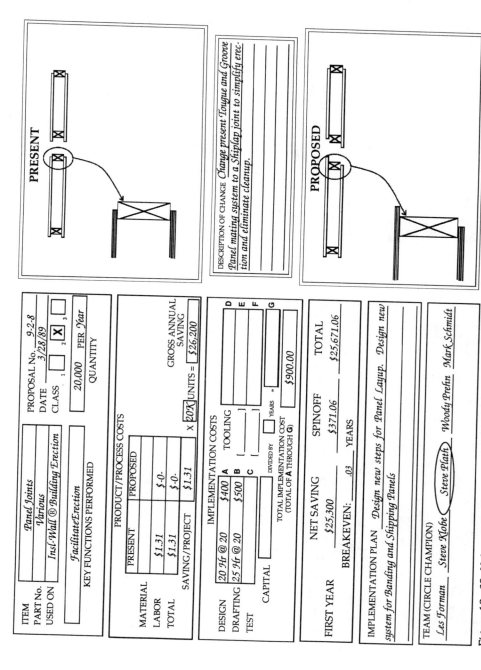

Figure 13–63. *Value analysis proposal 9-2-8 resulted in use of a shiplap joint for the panels; the gross saving was $26,200 per year.*

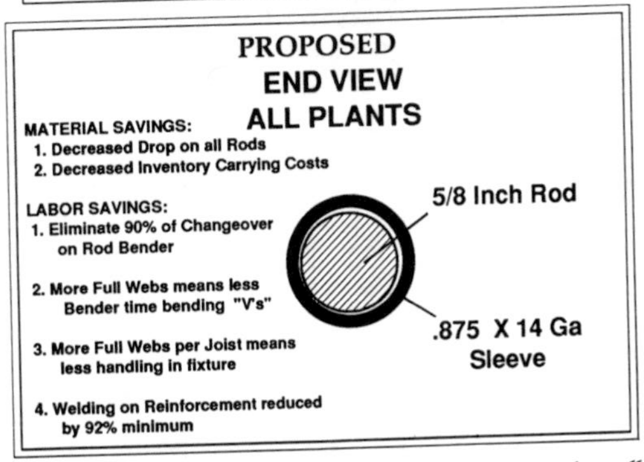

Figure 13–64. Value analysis proposal 3-1-5 designed a standard method of reinforcing the structure, for use by all Butler plants; this proposal improved appearance and reduced gross annual cost by $96,000.

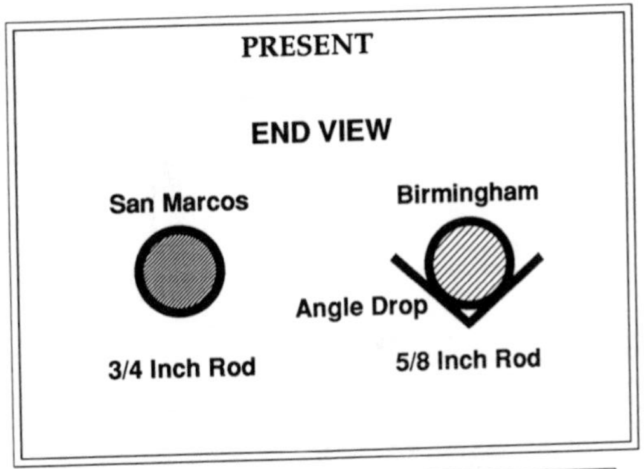

called Trenchduct®, a system that is installed in newly constructed office buildings to provide flexible electrical, telephone, and signal distribution.

The team developed a FAST diagram with 30 functions. Costing of the diagram revealed that the function "establish screed" required 11.8 percent of the total cost. The data from a focus panel of five users and eight Walker decision makers resulted in 42 likes and 8 dislikes. Three of the likes are shown in Figure 13–65.

All three of the elements relate directly to that function, which is essentially performed by adjusting the position of an angled "foot" that supports the power partition. The present method of adjustment is an elongated slot with a screw and nut. After adjustment, the foot is tack-welded to the power partition. Elements 5 and 36 show that the users feel the adjustment feature is important to them and that they are happy no partition-to-deck welding is required; element 8 shows that the users are not turned on by the present screw-and-nut approach.

The team identified "establish screed" as a value analysis target. The members applied their creativity and synthesis efforts to that function and prepared several proposals relating to the design of the adjustment feature. One is shown in Figure 13–66.

35. Restyled Oven Door
The Product: Restaurant oven/range
Lesson Learned: Where style controls, value analysis delivers style at minimum cost

The same geographically separated team that accomplished the stable stove-top spider redesign described in case study 25 mounted a major challenge to a top management fixation: The oven door was designed with an easily identified bulge, which many regarded as the very personification of the high quality of the product.

In the user focus panel, one of the company participants suggested that the bump be rated in terms of its importance to users. The focus panel data clearly indicated that the users generally regarded the feature as important to them. Indeed, six of the nine actual user/customers voted the bump as a 7 to 10 on the importance scale.

When the team members completed their microallocation of the $392 cost of the range to the functions on the FAST diagram, they discovered that function 7A, "improve

LIKES (Features or Characteristics)													
	1	2	3	4	5	6	7	8	9	10	11	12	13
5 Positive/Negative Prepour adjustment feature to accommodate differences in floor plane	8	8	8	8	3	9	9	8	10	9	10	3	8	MODE: 8R7
8 Screw vs Clip mechanism to establish temporary position of power partition prior to welding	5	8	3	8	9	2	2	1	3	3	9	6	3	3R8
36 No welding is required to attach to deck	5	10	10	10	10	2	8	9	8	9	8	9	9	9R8

Figure 13–65. An excerpt from the Walker focus panel shows three likes expressed by users/ customers, which directed the value analysis team toward the adjustment of the U-compartment.

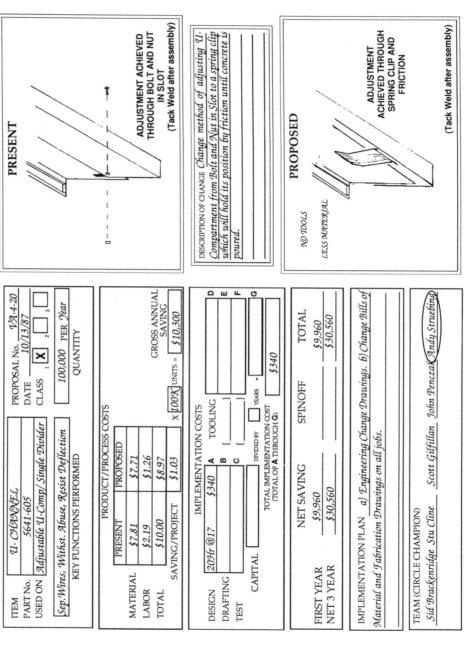

Figure 13–66. Value analysis proposal VA-4-20 replaced the bolt and nut formerly used in adjusting the U-compartment with a clip, for an annual gross saving of $10,300.

appearance," accounted for only 7.1 percent of the total. One of the contributors to this function cost was the $4.37 labor cost to form the bump in the door. One of the team members had long wanted to remove the bump, and this cost data strengthened his conviction. An intense team discussion followed. How could a new design satisfy the market need apparent from the focus panel data while eliminating the expensive labor involved in forming the bump?

The assistant plant manager was a team member. (He has been since promoted to plant manager.) His understanding of sheet metal forming led him to conceive, and later champion, the solution shown in Figure 13–67.

The embossed version is expected to satisfy the user desire for a visually attractive three-dimensional appearance. The cover will be formed in one press operation and requires no welding and grinding.

36. Sequencing System for Vault Alarm
The Product: Electronic circuit board
Lesson Learned: Value analysis works on electronic circuitry

A security systems manufacturer was faced with a massive onslaught of offshore competition. The company's two choices were to buy its electronic subassemblies from the Orient or reduce the cost of its PC boards by 60 percent!

Since it had installed expensive, high-volume PC board fabrication and assembly equipment, the company was inclined toward the cost-reduction solution. The 60 percent goal was daunting, but the vice president of manufacturing felt that such a dramatic goal could be realized through use of value analysis. As a young GE engineer, he had participated in a value analysis team that had reduced the cost of an electronic circuit by 73 percent while greatly improving both reliability and performance.

An outside consultant presented to the company president and his staff a plan for a two-team value analysis effort. After a spirited discussion, led by the vice president of manufacturing, the value analysis approach was given three months to solve the problem. All necessary company resources were to be dedicated to the effort.

One value analysis team was assigned to each of the two PC boards that together accounted for 86 percent of the electronic circuitry cost of the system. The value analysis consultant insisted that both teams attack the problem from the viewpoint of the entire security system, though each would concentrate its initial analysis on its assigned board. He was overruled by the vice president of R&D, whose value analysis experience was limited to the original GE style, in which the focus is on parts rather than on the purchased product.

Each team, therefore, created a FAST diagram for its assigned board. Figure 13–68 is the FAST diagram created by one team for its product; the signal sequencer PC board.

That team then allocated the $182.40 cost of the board to the FAST diagram. This 11-hour process forced the team to define the functions of each of the 876 line items of cost rigorously, developing a collective understanding of the reasons for the existence of each element. During the function-cost allocating process, the team made 32 entries in its Idea Bank notebook for use as triggers during the creativity phase. When the members posted their allocations to the FAST diagram, they found several surprises:

The five basic functions accounted for only 36 percent of the total cost.

The cost of the function "drive circuits" alone was 22 percent of the total board cost.

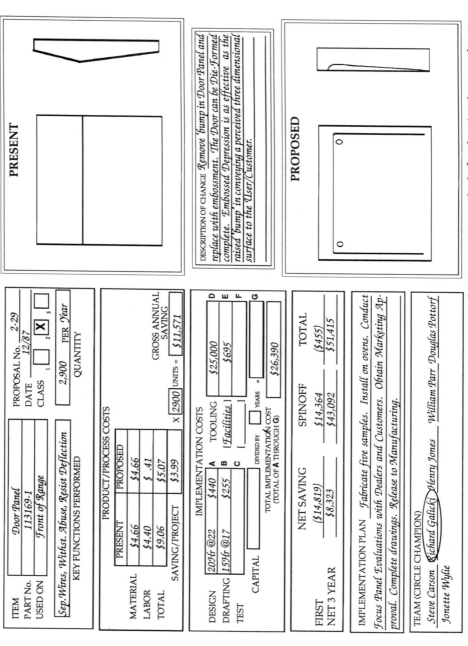

Figure 13–67. Value analysis proposal 2-29 produced an attractive but inexpensive method of embossing the oven door to maintain the perceived and desired three-dimensional effect.

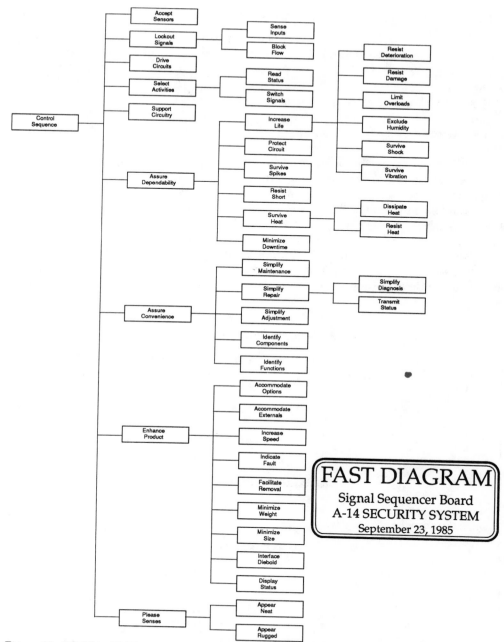

Figure 13–68. This FAST diagram of 33 independent functions describes all of the mechanical and electrical functions of the signal sequencer.

The other 44 percent of the cost was allocated to the function heading "assure dependability."

A user focus panel yielded 84 elements of user data, which the team also allocated to the functions of the FAST diagram. An excerpt from the focus panel report is shown in Figure 13–69.

The team allocated all four of these excerpted responses to functions under the general heading "assure dependability." The messages from the user/customers were clearly "Dependability is very important to us" and "We find it difficult to diagnose failures."

As they approached the analysis phase, the team members expressed frustration with the narrowness of their FAST diagram. They concluded that if the viewpoint of the diagram were broadened to cover the user-related functions of the entire security system, both function cost and function worth would better relate to reality.

In the analysis phase, the team reviewed the massive quantity of function cost and function worth data that they had developed during the first three days of team sessions and then decided that the functions "simplify diagnosis," "transmit status," and "drive circuits" were the most promising value analysis targets. Their creativity phase resulted in several hundred words and phrases triggered by these three functions. The synthesis

---

LIKES (Features or Characteristics)

	1	2	3	4	5	6	7	8	9	10	11	12	13	14	15	16	17	18	MODE
3 We experience a 20,000 hour MTBF on Sequencer Board	6	9	9	9	10	6	6	8	8	-	-	6	-	9	10	9	10	9	9R4
9 Board works in presence of noise and spikes	10	10	10	8	10	9	9	8	10	-	-	5	-	6	8	8	10	10	10R5
28 Board is tested to 200C ambient. Our ambient is 135C or less	9	8	9	10	9	8	9	9	9	-	8	5	-	10	9	9	6	9	9R5

DISLIKES (Faults or Complaints)

	1	2	3	4	5	6	7	8	9	10	11	12	13	14	15	16	17	18	MODE
32 It is difficult to diagnose circuit failures	9	9	8	8	8	10	8	8	9	9	-	8	-	10	9	8	8	9	8R3

PARTICIPANTS

COMPANY:

1 VP, Marketing
2 Director, Sales
3 Regional Sales Manager
4 Regional Sales Manager
5 Product Manager, Security Systems
6 Manager, Packaging Design
7 Manager, Electronic Design
8 VP, Research & Development

USERS/CUSTOMERS:

9 Maintenance Supervisor, Bank
10 President, Security Services Company
11 President, Department Store
12 Repair Service Technician
13 Director, Operations, Military Contractor
14 Maintenance Supervisor, Repair Service
15 Security Technician, Bank
16 SecurityTechnician, Retail Store
17 Security Technician, Warehousing Operator
18 SecurityTechnician, Courier Service

Figure 13–69. The three likes and one dislike expressed by the user/customer focus panel all relate to the single function group headed "assure dependability."

PRESENT

COMPONENT COUNT

DIP Chips	31
Capacitors	12
Resistors	16
Crystal	1
Power Drivers	4

DESCRIPTION OF CHANGE *Replace the four Power Drivers with one Power Driver and a four-port Power Switcher. Add leads to edge connector to feed signals to existing System Diagnostic Display. Restructure Digital Data Handling System eliminating Clock and taking advantage of K-14 chip.*

PROPOSED

COMPONENT COUNT

DIP Chips	21
Capacitors	8
Resistors	10
Crystal	0
Power Drivers	1
Power Switcher	1

ITEM	*Signal Sequencer Redesign*	PROPOSAL No. *1-3-6*
PART No.	1340-4210	DATE *9/23/85*
USED ON	*A-4 Security System*	CLASS 1 ☐ 2 ☐ 3 ☒

(Task) Control Sequence
KEY FUNCTIONS PERFORMED

1983 PER *Year*
QUANTITY

PRODUCT/PROCESS COSTS

	PRESENT	PROPOSED
MATERIAL	$158.80	$58.06
LABOR	$23.60	$18.52
TOTAL	$182.40	$76.58
SAVING/UNIT		$105.82

x *1983* UNITS = GROSS ANNUAL SAVING $20,984

IMPLEMENTATION COSTS

					TOOLING	
DESIGN	220 Hrs @22	$4,840	**A**			**D** $6,800
DRAFTING	30 Hrs @18	$540	**B** [ ]			**E** [ ]
TEST	85 Hrs @18	$1,530	**C** [ ]			**F** [ ]
CAPITAL						**G** $13,710

DIVIDED BY [ ] YEARS =
TOTAL IMPLEMENTATION COST (TOTAL OF **A** THROUGH **G**)

	NET SAVING	SPINOFF	TOTAL
FIRST YEAR	$7,274	$48,280	$55,554
NET 3 YEARS	$49,242	$144,840	$194,082

IMPLEMENTATION PLAN *Breadboard Circuit. Perform System Compatibility Tests. Design Circuit. Build 25 samples and put on test at Firnabank. Rewrite Sales and Service Documentation. Release to Manufacturing.*

TEAM (CIRCLE CHAMPION)
Chuck Smith *Fran Carlisle* *Frank Cartwright* *George Horn* *Ted Fowler*

Figure 13–70. *Value analysis proposal 1-3-6 involved a total redesign focused by the "dependability" emphasis. The component count was reduced, and the output driver circuits were restructured to take advantage of a power switcher. Cost savings were nearly $12,000, and reliability and performance improved.*

phase then opened the floodgates. The team grouped, elevated, and finally recorded 120 specific concepts for change. Each was investigated by a champion, and 38 proposals were presented to management. The essence of the proposals is shown in Figure 13–70. Implemented proposals represent a 68 percent reduction in board cost. More important, the new PC board represents a considerably better product in the eyes of the customer.

EPILOGUE: ESTABLISHING A VALUE ANALYSIS SYSTEM

ENVIRONMENT

Value analysis is powerful, and it is delicate.

Its full potential can be realized only in an environment that supports and nurtures it.

Establish an infrastructure in your organization that facilitates your value analysis operation.

The optimum structure must be customized to fit the objectives and traditions of the organization, but it will generally comprise the elements shown below:

Value council: a policy and review group led by the top operating manager

Guidelines: a policy and procedure document that defines responsibilities

Format: a standard form for submittal, approval, implementation, and audit

Audit: a formal, independent system under the direction of the comptroller

Reports: quarterly and annual, with an appended independent audit report

Throughout this book the emphasis is on principles that ensure the success of the value analysis system.

Listed below is the essence of 20 key principles. A value analysis system that fails to follow these principles will be seriously weakened.

20 PRINCIPLES

Management Considerations

1. Friend in court: A value analysis system cannot survive as a "revolution from below"; there must be a center of informed support in executive management.

2. Patience: The usual cost-reduction system pays off in weeks; a value analysis study strikes at the heart of the problem but may not pay off for months. Do not insist that your value analysis system pay off prematurely. You may be forcing it away from the sources of its great power: the job plan and the functional approach.

3. Low key: Publicizing the startup of a value analysis system does, indeed, cause a positive swing in attitude. Unfortunately, this only serves to accentuate an inevitable negative swoop. Remember—you are asking professionals to change the way they solve problems. Such a behavior change must be solidly based on favorable experience. Build your system slowly.

4. Recognize authority: Do not permit your value analysis system to circumvent the decision-making authority of anyone in your organization. Changes can be made only by those who have been duly authorized. Changes cannot be made by the value analysis system.

5. Build internally: Assign a motivated employee as the value analysis coordinator. Do not create a staffed value analysis department; value analysis does not take place in a value analysis department. It takes place in the exchanges among various elements of the organization. The job of a full-time value analysis coordinator is to guide teams of experts in the performance of value analysis.

Application Principles

6. Fit to organization: Your organization is unique. It is successful. Do not force it to change to accommodate the value analysis system. Force value analysis to adapt itself to your unique constraints.

7. Promise only the probable: Value analysis studies delight in predicting huge potential savings. Forego the thrill. You will implement only a portion of such savings; then you will have to explain how you "lost" the rest. Promise only savings that are probable.

8. Not simply motivation: Resist the temptation to accept the value analysis system just because "at least it will stimulate my people to consider cost." There are many fine ways to motivate employees. Value analysis is a problem-solving system. Use it to solve problems.

9. Don't train—solve: The original GE value analysis system did a lot of training of large groups in one-week, 40-hour programs. They didn't work very well, but they survived for 12 years because they saved some money, they were fun, and GE put $3 million into them. Concentrate on problem solving. Then you'll get even better training results.

10. Teach by involvement: Value analysis cannot be taught by lecture. It cannot be learned by reading a book. No one has ever understood the unique power of the process without hands-on experience. Teach it by giving a high-level team the opportunity to experience a dramatic success.

11. Key man: The team must include the key man—the most technically knowledgeable person available. Ideally, he or she is the project engineer on the product under study.

12. Preparatory work: Relieve the team of unmotivating drudgery. Develop complete and detailed data on function and cost before the team meets.

13. Convince everyone: Every participant in a value analysis team must become convinced that value analysis is the most effective system available for solving function-cost problems. An unconvinced participant becomes a localized site of infection. This is

fatal to the horizontal interaction that is essential to the effectiveness of the value analysis system.

14. Function analysis: If function is not analyzed, it is not value analysis. This is the most vital element that gives value analysis its great power to cut to the heart of the problem. Modern value analysis uses the FAST diagram to guide its function analysis.

15. Function-cost: Rigorous, detailed costing of each of the functions on the FAST diagram is an essential step to the true understanding of the project under study.

16. Function-attitude: It is essential to relate user attitude in equally rigorous detail to the functions on the FAST diagram, to ensure that the first emphasis of value analysis remains on fulfilling user/customer needs and wants.

17. Create by function: Apply creative problem solving techniques to the abstract function rather than to the concrete problem. This ensures that your solutions will not be unduly constrained by precedent.

18. Aim your creativity: Do not waste your creative effort on problems that do not need solving. Focus your effort only on functions with low user acceptance and/or high cost.

19. Implementation: As the first step in the problem-solving process, develop a plan for the implementation of any possible solution. Update that plan as the workshop continues. The name of the game is implementation.

20. Champion principle: Recognize the power of individual ownership. Do not present a value analysis proposal unless a value team member has agreed to become its champion.

APPENDIX A: POLICY AND PROCEDURE

THE CORPORATE VALUE ANALYSIS AND COST-IMPROVEMENT SYSTEM [1]

1 Purpose

To establish corporate policy and procedure relating to the corporate value analysis and cost-improvement system.

2 Applicability

All plants, divisions, groups, and subsidiaries of the corporation.

3 Policy Statement

The value analysis and cost-improvement system is a fundamental part of the corporation's operating philosophy, and is intended to generate a continuing flow of cost-improvement savings and product/process improvement. The system includes the elements of value analysis, traditional cost improvement, scrap reduction. Delivery cost reduction, and operations or methods improvement. Each location is expected to make an appropriate contribution to increased profitability.

4 Project Types

The following definitions are offered as a guide to categorizing projects properly within the value analysis and cost-improvement system. In applying these definitions to partic-

---

[1]The document has been adapted from the control document used by Philips Industries of Dayton, Ohio. It originated with the late Langston C. Schaefer, of Emerson Electric and Thomas Cook Associates, and was modified by Philips to meet that company's unique constraints.

ular situations, it is important to remember that only those projects properly arising from a formal value analysis workshop study are to be categorized as value analysis projects. Any other project must be placed in another category best describing its nature.

4A Value Analysis Project

Any cost-improvement or product-improvement project that is identified and reported during a formal value analysis workshop study is considered a value analysis project. Routine functions such as securing a favorable purchase price or developing a new product do not fit the value analysis category. A value analysis workshop study may, of course, be instituted to consider any area of cost, and is not restricted to the consideration of product cost only; in those cases, projects arising from such value analysis workshop studies would be identified as value analysis projects.

4B Scrap-Reduction Project

This category includes any project arising outside a formal value analysis workshop study and aimed specifically at reducing scrap, whether it be a saving in the cost of material, the cost of manufacturing, or the cost of handling scrap. Such projects might include higher recovery values for unavoidable scrap, better storage facilities to reduce in-process handling losses or tooling, or changes in methods to reduce losses in manufacturing scrap.

4C Delivery Cost-Reduction Project

This includes any project arising outside a formal value analysis workshop study and aimed specifically at reducing the cost of delivering goods to customers. Such projects might include the use of more efficient equipment, resulting in a lower overall cost; the use of larger trailers, to deliver more goods per mile driven; or the obtaining of increased backhaul revenues.

4D Operations- or Methods-Improvement Project

This category includes any project arising outside a formal value analysis workshop study and aimed at reducing the cost of a manufacturing function or a manufacturing support function. Examples of such projects are rearrangement of an assembly line; a change to more efficient manufacturing machinery or tooling, fixtures, or jigs; and relocation of parts storage to lower in-house handling costs.

4E Cost-Improvement Project

This category covers any project arising outside a formal value analysis workshop study and not clearly fitting into any of the foregoing categories. Examples could include reduction of energy costs; the elimination of a task or function as unnecessary to the achievement of established goals; a management project aimed at reducing insurance costs by decreasing the occurrence of on-the-job accidents; changes in paperwork systems to reduce the cost of printed forms or reduce document preparation and processing; and negotiation of lower material prices.

5 Project Savings Calculation Procedure

5A Define Costs

5A1 Implementation Costs

This refers to any one-time noncapital cost that must be incurred in order to implement any type of value analysis or cost-improvement project. Implementation costs normally include engineering design, drafting and testing time (except what will be capitalized as part of an asset), tooling or facilities modifications of an expense nature, and the first-year depreciation cost of a capital asset needed to implement the project. Useful asset lives consistent with those assigned in corporate fixed asset accounting should be used. Implementation costs would not include the base cost of any capital asset acquired in order to implement the value analysis or cost-improvement project.

5A2 Direct Material

Direct material refers to all material that forms an integral part of the finished product and is directly included in calculating the cost of the product per the bill of materials. Material savings are defined as the difference between the direct material amounts for the original product and the revised product.

5A3 Material Overhead

Material overhead is defined as the cost of indirect labor or indirect material that is incurred in ordering, expediting, receiving, testing, and storing purchased materials. Material overhead is usually expressed as a rate, either in terms of dollars per base unit ($/$ of purchased material) or as a percentage of base unit (% of purchased material $).

5A4 Direct Labor

Direct labor is the labor expended immediately upon the materials constituting the finished product. Direct labor may also be termed productive labor. Labor savings are defined as the difference between the direct labor costs for the original product and the revised product.

5A5 Manufacturing Overhead

Manufacturing overhead is defined as the cost of indirect materials, indirect labor, and all other manufacturing costs that cannot be directly charged to specific units or products, but instead are allocated to a product line. Manufacturing overhead is usually expressed as a rate, either in terms of dollars per base unit ($/direct labor hour) or as a percentage of a base unit (% of direct labor $). Manufacturing overhead is further subdivided into four categories: variable, semivariable, fixed, and fringe-only.

5A5a Variable Manufacturing Overhead

Variable overhead is the total of the overhead items that vary directly with production levels. Examples of variable items include indirect material, indirect labor, shop supplies, small tools, payroll taxes and fringes, receiving department expenses, shipping department expenses, and overtime premium pay.

5A5b Semivariable Manufacturing Overhead

Some expenses vary directly with production, but not in direct proportion to production volume. These are known as semivariable, and contain both fixed and variable elements. Examples include production supervision expense, inspection and quality assurance expenses, and maintenance and repair of machinery and plant equipment.

Unless it can be demonstrated that semivariable overhead items are affected by a value analysis or cost-improvement project, they are omitted from any savings calculation.

5A5c Fixed Manufacturing Overhead

Fixed overhead is the total of overhead items that do not vary directly with production levels. These are sometimes known as period costs, since they relate more closely to accounting periods than production levels. Examples include executive salaries, depreciation, real and personal property taxes, insurance (property and liability), rent, and repair and maintenance of building and grounds.

These overhead items are not included in savings calculations unless it can be clearly demonstrated that there has been a reduction in cost.

5A5d Fringe Only

Fringe is the total of the overhead items that vary directly with direct labor. Examples of fringe items include payroll taxes, employee insurance, employee pension payments, workers' compensation payments, and vacation and sick leaves.

Overhead rates are calculated using the form shown in Figure A–1. Rates may be expressed in relation to direct material or direct labor, whichever is most applicable. The overhead rate worksheet has some specific items listed as components of variable overhead. Space is provided for the insertion of additional relevant items. Only variable manufacturing expense items are to be included in the calculation, excepting certain semivariable expense items that can be demonstrated to apply. No fixed-manufacturing expense items may be included unless it can be clearly shown that there will be a reduction in that cost and this calculation shall be shown separately. An overhead rate should be established for each project, including as part of overhead only those overhead items that the project affects. All rate calculations should be made using year-to-date financial information.

PLANT or LOCATION	PROJECT No.		DATE

[All amounts are Year-to-Date]

DIRECT MATERIAL $		DIRECT LABOR $	
OVERHEAD AS A % OF MAT'L	%	OVERHEAD AS % OF DIRECT LAB	%
OVERHEAD ITEM		**OVERHEAD ITEM**	
INDIRECT PAYROLL		UNIFORM RENTAL & EXPENSE	
PREMIUM PAY (PRODUCTION)			
INDIRECT MATERIAL			
SMALL TOOLS AND SHOP SUPPLIES			
DIE REPAIR AND MAINTENANCE			
RECEIVING PAYROLL			
RECEIVING PREMIUM PAY			
RECEIVING SUPPLIES			
SHIPPING PAYROLL			
SHIPPING PREMIUM PAY			
SHIPPING SUPPLIES			
P/R TAXES & FRINGES (MFG)			
EQUIPMENT RENTAL (MFG)			
EQUIPMENT REPAIR (MFG)			
CUSTOMER CLAIMS AND WARRANTY			
ELECTRIC POWER			
INDUSTRIAL ENGNG PAYROLL			
INDUSTRIAL ENGNG PREMIUM PAY			
IND'L ENGNG P/R TAX + FRINGE			
INDUSTRIAL ENGINEERING SUPPLIES			
TOOL MAINTENANCE PAYROLL			
TOOL MAINTENANCE PREMIUM PAY			
TOOL MAINT P/R TAX + FRINGE			
TOOL MAINTENANCE SUPPLIES			
MFG ENGNG & QC PAYROLL			
MFG ENGNG & QC PREMIUM PAY			
MFG ENGNG & QC P/R TAX + FRINGE			
MFG ENGNG & QC SUPPLIES			
MATERIALS MGNT PAYROLL			
MATERIALS MGNT PREMIUM PAY			
MATERIALS MGNT P/R TAX + FRINGE			
MATERIALS MGNT SUPPLIES			
TOTAL		TOTAL	

Figure A–1. The overhead calculation form is a worksheet used for calculating the effect of a value analysis or cost-improvement project on overhead accounts.

5A6 Delivery Costs

Delivery costs encompass those expenses incurred to get the product from the point of manufacture to the customer. Included are costs to maintain and operate any fleet of company trucks, as well as charges for delivery of product by outside carriers. Delivery costs include only outbound freight. Inbound freight is charged either to inventory or to some other applicable expense category.

5A7 Selling, General & Administrative Costs

Selling, General & Administrative (SG&A) expenses are not manufacturing in nature, but relate to costs associated with selling the product, costs in administering the overall business from an executive level, and other costs not classified elsewhere.

5B Volume Increases

Volume increases that are forecast as a result of price reduction or other marketing efforts are not used as a basis for calculating savings. These increases are a result of management decisions outside the value analysis and cost-improvement effort. Also, the costs to accomplish these increases are not charged against the savings. Volume increases that result from increased customer demand based on product improvements are to be accounted for within the value analysis and cost-improvement effort. The basis for any such estimate is a rigorous analysis by the manager of marketing, with approval of the conclusions by the responsible assistant comptroller.

5C Spin-Off Savings

These are savings arising from the completion of a value analysis workshop study that can be easily incorporated into another, separate product. In order to receive credit for the additional savings, the follow-on project must be assigned a separate project number, and the accompanying value analysis proposal must be noted as being based on "spin-off saving from Project   ." The new project number should also be noted on the original proposal form.

5D Savings Calculation Procedure

Project savings should be evidenced by lower out-of-pocket expenditures, not merely a reallocation of costs. For this reason, fixed-overhead items such as depreciation, real and personal property taxes, executive salaries, and so on are not included in the savings calculation unless it can be clearly shown that there is a net reduction in these costs. Semivariable overhead items may be included to the extent that they are affected by the project (for example, if it can be shown that fewer foremen will be required due to direct labor reductions, the salaries of those foremen may be included as savings). Where a semivariable overhead cost saving or a material overhead cost saving is realized by a value analysis or cost-improvement proposal, enter the specific accounts and amounts in a value analysis and cost-improvement overhead rate worksheet. The worksheet is attached to the value analysis or cost-improvement worksheet.

6 System Implementation

6A Overview

Annual dollar savings targets per fiscal year are set as a percentage of each division's approved sales budget. Of this total, a portion is committed to the value analysis system, with the balance coming from the other cost-improvement systems. Operating management determines where to focus efforts among these other areas. The overall cost-improvement goal will be 5 percent of the approved sales budget, with the value analysis goal equaling 3 percent of sales or 60 percent of the overall goal.

The major emphasis of the system is on the identification and subsequent implementation of projects.

In measuring results, carryover savings from prior-year projects count toward achievement of the goal. If a large item of carryover develops at the end of a year and was not included in the goal setting, it does not relieve a division of the obligation to maintain a strong and active value analysis and cost-improvement system.

Both planning and monthly reporting for the operating divisions are carried out in two forms, one a project list (Fig. A–2), the other a summary of overall progress (Fig. A–3)

During preparation of the business plan, each division should schedule at least three value analysis studies in order to ensure an ongoing commitment to this vital system. While a division may schedule more or fewer than three studies, it is expected that every reasonable effort will be made to hold the value analysis studies as scheduled. The following sections discuss the general procedures under which this system operates, and provide specific instructions for completing the two key forms that are used for tracking and reporting purposes.

6B Preparation of Annual Business Plan

6B1 Value Analysis and Cost-Improvement Project List

All projects that have been identified are listed on a value analysis and cost-improvement project list (Fig. A–2), indicating the project number, description, coordinator, and planned/estimated results.

A separate project list should be filled out for each type of value analysis or cost-improvement project, with each page marked to indicate the type of projects listed on that page.

6B2 Value Analysis and Cost-Improvement Summary

After completing the various project lists, the division's cost-improvement program summary (Fig. A–3) is filled out.

On this form is shown the following data:

The overall dollar goal (for example, 5 percent of sales), divided between value analysis and other cost-improvement systems.

The prior-year carryover from each cost saving area.

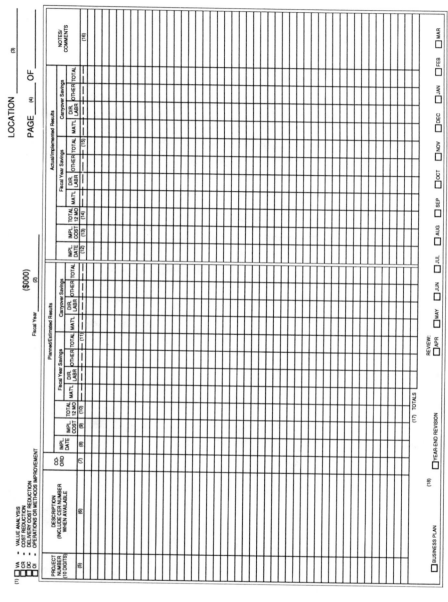

Figure A–2. *This is a columnar listing of all value analysis and cost-improvement projects; it is issued at the time of the business plan preparation, and at specified times and in specified categories thereafter.*

---

(1)   DIVISION

FY 19____ COST IMPROVEMENT GOALS: GOAL ($000)

    VALUE ANALYSIS   _____ % OF SALES $ (3)   _____

    OTHER SOURCES   _____ % OF SALES $ (4)   _____

        TOTAL GOAL   _____ % OF SALES $ (2)   _____

		NEW SAVINGS REQUIRED		
	CARRYOVER FROM FY 19____ (5)	* IDENTIFIED (8)	NOT IDENTIFIED (9)	TOTAL (6)
VA = VALUE ANALYSIS				(7)
CR = COST REDUCTION				
SR = SCRAP REDUCTION				
DC = OPERATIONS OR METHODS IMPROVEMENT				
TOTALS (10)				
% OF TOTAL (11)				100.0

* GUIDELINE TO IDENTIFICATION OF SAVINGS

      When Business Plan is submitted   60% to 70%
      By April 1   85% to 90%
      By July 1   90% to 95%
      By October 1   100%

SCHEDULED VALUE ANALYSIS PROGRAMS:

    PLANNED DATE: (12) _____   _____   _____   _____

    ACTUAL DATE: (13) _____   _____   _____   _____

    $ SAVINGS IDENTIFIED   (14) _____   _____   _____   _____

PREPARED FOR: ☐ BUSINESS PLAN

              ☐ YEAR-END REVISION

              ☐ REVIEW FOR MONTH OF [_____]

Figure A–3. The project summary form summarizes the effect on the business unit's performance of the value analysis and cost-improvement projects listed on the project list.

The new savings already identified (taken from the previously completed project lists).

The amount of savings not identified.

The total saving to be realized from each cost saving area.

A list of scheduled value analysis studies for the plan year—at least three studies should be scheduled.

Since this form is completed prior to the end of the current fiscal year, some figures will be subject to adjustment—what will not change are the total value analysis and cost-improvement dollar goal and the minimum value analysis savings goal.
What will change are the following:

Estimated carryover, based on actual accomplishments for the current year.

New savings required (based on a change in carryover).

Possibly the mix among the several savings areas (except value analysis).

The project list, which should be brought to a suitable beginning point for tracking throughout the new fiscal year.

It is also recognized that priorities may change during the year as unanticipated savings opportunities arise. Therefore, the mix of savings (except value analysis) may further change as the year progresses.

6B3 File Locations

Place the completed project summary and cost-improvement project lists in the cost-improvement system section of the business plan book.
Enter the total savings goal on the fiscal year 19XX business plan analysis form in order to arrive at the total FY19––business plan objective.

6B4 Assignment of Project Numbers

In order to count toward achievement of the committed value analysis and cost-improvement goal, each project must be assigned a number that will facilitate identification and tracking. This consists of 12 characters, broken down as follows:

> **1 and 2.** Fiscal year in which project number was assigned
>
> **3 and 4.** Plant or location number (2 digits)
>
> **5 and 6.** Type of project
> VA = value analysis
> CR = cost reduction
> SR = scrap reduction
> DC = delivery cost improvement
> OI = operations or methods improvement
>
> **7.** dash
>
> **8.** Class of proposal or project
> 1 = clean, an immediate "go"; payback normally less than 12 months
> 2 = looks good, but requires some testing or market approval; payback normally less than 24 months

3 = major change; may involve large capital investment, product redesign, or policy change

9. dash

10 to 12. sequential number of project

For example, project no. 8709VA–1–123 was assigned in fiscal year 1987 at the plant identified as 09; it is a value analysis project, class 1, and its specific number is 123.

Any project that remains unimplemented at the end of a fiscal year, but is still viable, is carried forward to the next year's list under its original number.

6C Reporting Dates

6C1 Annual Business Plan

Initially submit project lists (multiple pages should be used as needed) and project summaries as the cost improvement system section of the business plan book.

6C2 Fiscal Year's End

Revised project lists and project summaries (adjusting for the actual carryover as of fiscal year end) should be submitted as a basis for tracking throughout the new fiscal year. The total dollar savings commitment contained in the business plan does not change, and any reduction in carryover has to be replaced with additional new savings.

6C3 Monthly Reporting

For the divisions, the main control and reporting document is the value analysis and cost-improvement project list (Fig. A–2).

During the year, implementation of identified projects is recorded on project list forms as actual results (when they are implemented).

Additionally identified projects are added to the list; those that prove unfeasible will be noted as such, but not removed from the list. Savings reported as actual are, in fact, estimated savings. The division coordinator monitors these estimated savings and revises the estimates if there is a substantial change in the planned saving.

At the end of each year, every item on the project list must be disposed of by either dropping the project as unfeasible or, in the case of an unimplemented but still viable project, placing it on the next year's project list, retaining the original project number.

Each division submits a copy of its project lists and project summary at the end of each month (by the second workday of the following month) to the corporate manager for value analysis, corporate manufacturing services department, and assistant corporate comptroller.

From the lists submitted, the corporate manager for value analysis prepares a value analysis and cost-improvement system performance summary (Fig. A–4) for each division, as a means of keeping division presidents and senior corporate management abreast of each division's progress toward its savings goal.

This report shows the following:

MONTH									DIVISION

		IDENTIFICATION OF NEW FY1988 SAVINGS					IMPLEMENTATION OF NEW FY1988 SAVINGS		
*	CARRYOVER FROM FY 19___	** IDENT	% OF GOAL	NOT IDENT	% OF GOAL	TOTAL GOAL	FY 19___ IMPACT	% OF GOAL	TO FY 19___	TOTAL
VA										
CR										
SR										
DC										
OI										
TOTALS										

```
*    VA   =    VALUE ANALYSIS
     C R  =    COST REDUCTION
     S R  =    SCRAP REDUCTION
     DC   =    DELIVERY COST REDUCTION
     O I  =    OPERATIONS OR METHODS IMPROVEMENT

**   TARGETS FOR IDENTIFICATION OF SAVINGS
          BY APRIL 1        60%   to    75%
          BY JULY 1         85%   to    90%
          BY OCTOBER 1            100%
```

PROJECTS SCHEDULED BUT NOT IMPLEMENTED

PROJECT #	PROJECT DESCRIPTION	COORD	SCHEDULE	ANNUAL SAVINGS

VALUE ANALYSIS SESSIONS

	1	2	3	4
PLANNED DATES				
ACTUAL DATES				
SAVINGS IDENTIFIED ($000)				

Figure A–4. The system performance summary form allows corporate management to evaluate the performance of each business unit in the value analysis and cost-improvement program.

The final adjusted prior-year carryover figure (which will remain static unless some unusual circumstance justifies a change).

Progress toward identification of the required new fiscal-year savings.

Progress toward implementation of identified projects.

Scheduled but unimplemented projects.

The results of value analysis studies held, with savings identified.

From the division performance summaries, a consolidated value analysis and cost-improvement system performance summary is prepared for senior corporate management (Fig. A–5).

MONTH:	**CONSOLIDATED SUMMARY**									
		IDENTIFICATION OF NEW 1988 SAVINGS					IMPLEMENTATION OF NEW FY1988 SAVINGS			
DIV	CARRYOVER FROM FY 19____	** IDENT	% OF GOAL	NOT IDENT	% OF GOAL	TOTAL GOAL	FY 19____ IMPACT	% OF GOAL	TO FY 19____	TOTAL

** TARGETS FOR IDENTIFICATION OF SAVINGS

BY APRIL 1	60%	to	75%
BY JULY 1	85%	to	90%
BY OCTOBER 1	100%		

Figure A–5. The consolidated performance summary, prepared for senior corporate management, compares plans and accomplishments of each division.

This report shows the same basic information as the performance summary, but in a condensed form.

6D Records Retention

Each division's value analysis and cost-improvement coordinator keeps a file of all cost-improvement projects, with supporting data. Data on file for each project should include (but is not limited to) the following:

The original project proposal worksheet

Worksheets or other materials to support the total proposed cost improvement, including information on volume

Sales projections for cost-improved products as sales increases, factored into the savings calculation, as well as any planned price reductions or increases

Implementation cost estimates

Vendor quotes and correspondence

Any other data, calculations, or correspondence that supports project costs, savings, spin-off effects, and implementation dates and costs.

6E Coordinators

6E1 Corporate

The corporate manager for value analysis has responsibility for overall direction and implementation of value analysis and cost improvement at the division level. This includes, but is not limited to, the following:

Selection of qualified division coordinators in cooperation with division management

Leading formal value analysis studies

Preparation of accurate documentation for value analysis projects

Timely follow-up and implementation of value analysis and cost-improvement projects

Timely and accurate reporting of those projects and the results

Division follow-up to ensure compliance with the total value analysis and cost-improvement system.

6E2 Division

All divisions appoint a value analysis and cost-improvement coordinator to be responsible for reporting project status to corporate management as required by policy. This responsibility includes review and audit of actual savings calculation upon implementation, insuring that implementation has in fact occurred, and expediting projects as necessary.

6F Instructions for completing the Project List and Project Summary

6F1 Completion of Value Analysis and Cost-Improvement Project List (Fig. A–2)

This form is used for both planning and reporting purposes.

As part of the annual business plan, this form lists all projects identified at that time and part of the division's committed dollar savings goal.

At this time, planned/estimated results are shown for each project listed. Savings are estimated for the 12-month period following the planned date of implementation. Estimated savings are separated into material, direct labor, and other cost components. The 12-month savings are allocated to the fiscal years that they will affect, based on the planned implementation date.

The forms submitted in the annual business plan are updated at year's-end in order to define more accurately the amount of carryover from the year just ended and provide a starting point for tracking new fiscal-year results.

For reporting purposes, implementation of listed projects is shown in the actual results columns. Here again, actual savings are estimated for a 12-month period, then separated into fiscal-year and carryover components, as well as material, direct labor, and other cost components. The division coordinator monitors these estimated savings and revises them if there is a substantial change from the planned savings.

Filling Out the Project List

Numbers refer to those on Figure A–2. Items 1 through 11 are completed during the preparation of the annual business plan for all projects identified at that time.

1. *Type of projects:* Indicate the type of projects on this list. (As a general rule, each type should be kept on a separate page.)

2. *Fiscal year:* Indicate the fiscal year.

3. *Location:* Identify the reporting division. (If desirable, lists may be maintained by the plant; however, only a division list is required.)

4. *Page:* Where a division has more than one page, all pages are numbered. Add numbered pages to the series as they are created.

5. *Project number:* Record the project number in accord with the coding instructions in section 6B4 of these instructions.

6. *Description:* This is a brief description or name of project. Include CER number when available.

7. *Project coordinator:* This is the person directly responsible for carrying out the project. In the case of a value analysis project, the project coordinator is initially the champion.

8. *Implementation date:* The month and year of the planned implementation date are inserted.

9. *Implementation cost:* Insert the estimated cost from the cost-improvement or value analysis proposal form.

10. *Total 12-month saving:* Insert the estimated 12-month savings from the cost-improvement or value analysis proposal form. This amount is the gross saving less implementation cost.

11. *Total:* This is the estimated saving impact, based on the planned implementation date and the expected savings flow (considering sales volume and other appropriate factors) over the 12-month period following implementation. Estimate the savings benefit within the fiscal year of implementation and the savings that will be realized in the following fiscal year. Show each in terms of material, direct labor, and other cost components. The total savings impact must equal the 12-month savings total.

12. *Implementation date:* The actual implementation date goes here—the month in which the project is actually put into effect (in other words, made part of the production cycle).

13. *Implementation cost:* Insert the actual amount.

14. *Total 12-month savings:* This amount is the actual 12-month savings from the implemented project—gross savings less implementation cost.

15. *Total:* This is the actual savings impact of the implemented project, based on the actual savings flow (considering volume and other appropriate factors) over the 12-month period following implementation. Estimate the savings benefit within the fiscal year of implementation and the savings benefit that will be realized in the following year in terms of material, direct labor, and other cost components. The total savings impact must equal the 12-month savings total.

16. *Notes/comments:* This section may be used for any appropriate comments regarding a project. If a project is dropped or not implemented on schedule, a brief explanation must be given. (Refer the explanation to an attached sheet if necessary.) At fiscal year's end, all unimplemented projects must be designated as dropped or carried over. The latter indicates that the project is being placed on the following year's list. The original project number will be retained to identify that the project has been carried over. Upon implementation of the project, savings will count toward the new fiscal year's goal.

17. *Totals:* This represents the totals for all items on the list, whether in the evaluation stage or already implemented.

18. *Purpose:* Indicate the reason for the preparation of this issue of the project list.

6F2 Completion of Value Analysis and Cost-Improvement Project Summary (Fig. A–3)

This form is used for planning and reporting purposes and is intended to summarize the division's overall value analysis and cost-improvement system goal, the areas from which savings are expected to be generated, and to provide a monthly summary of progress toward those goals.

At the end of the current fiscal year, the carryover figures are adjusted for actual results through March, and the identified and unidentified portions of the required new savings are updated. This updated project summary (which does not alter the total dollar savings commitment) and the supporting project lists are the starting point for tracking performance throughout the new fiscal year.

Each month a project summary is submitted, along with the updated project lists (by the second workday of the succeeding month). The monthly project summary describes the current status of the division's value analysis and cost-improvement system, as well as indicates the results of recent value analysis workshop studies.

Filling Out the Project Summary

Numbers refer to those on Figure A–3.

1. *Division:* Insert the division name.

2. *Total goal:* This is the total value analysis and cost-improvement system savings goal for the year (for example, 5 percent of the approved sales budget for the year).

3. *Value analysis:* This is the value analysis savings goal for the year (for example, 3 percent of the approved sales budget for the year).

4. *Other sources:* These are the savings goals for projects other than value analysis; it equals the difference between the total goal (2) and the value analysis goal (3).

5. *Carryover:* This is from the prior fiscal year, estimated when preparing the annual business plan in order to identify the amount of new savings required. It is revised at year's end based on actual activity for the full fiscal year.

6. *Total (value analysis):* This is the same as 3.

7. *Total (categories other than value analysis):* This is the same as 4, but divided into more specific categories. Specific amounts in each category may be revised as the year progresses.

8. *Identified:* These are projects identified as of planning time. Identified projects must be backed up by completed value analysis proposal forms.

9. *Not identified:* This is the amount of savings yet to be identified—the total (6) less carryover (5) and identified (8).

10. *Totals:* These are the column totals.

11. *% of total:* The percentage of totals for the carryover, identified, and not identified columns.

12. *Planned date:* Insert the month and year of scheduled value analysis workshop studies. At least three sessions should be scheduled as part of the annual business plan.

13. *Actual date:* Insert the month and year of value analysis sessions actually held.

14. *$ savings identified:* These are the dollar savings resulting from each value analysis workshop study held. Savings should be reported for all proposals generated during each value analysis workshop study.

15. *Prepared for:* Indicate the purpose of the current submittal of the project summary.

APPENDIX B: A COLLECTION OF REQUIRED FORMS AND CHECKLISTS

The following forms and checklists supplement the discussions and analysis found in this book. While they may be modified according to individual needs, most readers will find the basic formats, as presented here, to be both useful and appropriate to their value analysis situations.

PLANNING GUIDE (Weeks prior to 1st team meeting)	SCHEDULED COMPLETION DATE	ACTIONS TO BE TAKEN
SIX	_____	Workshop dates and hours scheduled Projects Selected Representative appointed to liaison with Workshop Staff
FIVE	_____	Team members identified Focus Panel arrangements complete
FOUR	_____	Team members notified by letter Collection of Project Cost Data begun Workshop site reserved
THREE	_____	Supplier/Expert arrangements begun Attendance of Top Management at first Workshop session confirmed
TWO	_____	Conduct Focus Panel or complete User interviews Workshop equipment and logistics complete Project technical and cost data complete Reminder memo with agenda sent to each participant Cost Data requirements complete
ONE	_____	Supplier/Expert arrangements complete (Including head count for lunch) List of invited Experts/Suppliers prepared List of participants with titles and duties finalized

Figure B–1. Workshop Coordinator's Preparation Checklist

1. Reserve off-site meeting room. Arrange for a message handling service outside the workshop room so messages can be held until break.

2. Overhead Projector. *Transmissive* not *reflective* type (bulb in base, not in head). Spare bulb.

3. Projection screen. Minimum size 5 feet by 5 feet.

4. "VHS" Video Cassette Player with Monitors for teams. For use on Synthesis-day only: (_____).

5. Twenty--five--foot multiple outlet extension.

6. A table for each team. (Round tables for six persons are preferred.) Also some six-- or eight--foot reference tables for staff use and vendor displays.

7. Arrange for a photocopier to be available to team members.

8. Luncheon should be arranged for the participants. Limit to 45 minutes if possible.

9. A group lunch with the teams and the Experts should be arranged for Synthesis-day.

10. Rolls and coffee should be on hand in the morning, 30 minutes before starting time and at morning and afternoon breaks. Soft drinks are also suggested for the afternoon break.

11. Safety pin type Lapel Name Tags. Both names. **Bold** printing.

12. Blank name tags for visitors and Experts/Vendors.

13. One 28--inch--by--32--inch Flip Chart Easel for each team (with extra pads, preferably with grids).

14. If applicable, have a basic set of tools available.

THE FOLLOWING ITEMS SHOULD BE MADE AVAILABLE TO THE WORKSHOP STAFF ON OR BEFORE THE FIRST WORKSHOP SESSION:

1. Three copies of a complete list of participants by team, job title and area of specialization.

2. Copies for each team member of a list of the Experts/Vendors who have been invited to participate on the Synthesis day, showing their organizations and areas of specialization.

Figure B–2. Equipment and Logistics List

The purpose of this questionnaire is to secure your opinions as to the success of this Value Analysis Study. The results of this questionnaire will be used to optimize the Value Analysis System

1.   What is your overall opinion of this Study? Please place an X in the appropriate box.

EXCELLENT	VERY GOOD	GOOD	FAIR	POOR

20 19 18 17   16 15 14 13   12 11   10   9   8   7   6   5   4 3   2   1

2.   Why did you rate it this way? _____

_____

2.   Was there anything (else) about the Study that was unsatisfactory? ☐ Yes ☐ No
      If yes, what was it? _____

_____

_____

3.   On the whole, would you say this Study gave you what you expected?

      ☐ As Expected   ☐ Not as well as expected   ☐ Better

4.   How much room for improvement would you say is needed in the Study?

      ☐ A Whole Lot   ☐ A Fair Amount   ☐ Only a Little   ☐ None

5.   If a good friend of yours who has a job like yours asked if he should attend a similar Study, what would you tell him?

      ☐ Attend   ☐ Don't Know   ☐ Don't Attend

6.   Thinking over your entire experience, and assuming you had free choice all over again to participate in this Study, would you have participated?

      ☐ Yes   ☐ No   ☐ Don't Know

7.   What did you like *most* or *least* about this Study? _____

_____

_____

                          Signature (optional) _____

Figure B–3. Program Evaluation Questionnaire Form

_____ Secure from the Value Analysis Workshop Coordinator a precise definition of the projects under study and the number of teams working on each project. If more than one team is assigned to the same project, secure a precise definition of the criteria for segregating the project between teams.

_____ Order one set of Drawings and Specifications for each team.

_____ Work with Finance to determine the Variable Burden Rate to be used for direct labor.

_____ Deliver to the Value Analysis Workshop Coordinator a sample of the proposed Product or Process cost data format before _____. (Critical)

_____ Order cost data and Process Routing data for each team.

_____ When Drawings, Specifications and Costs are delivered, separate them by team based on the segregation criteria.

_____ Collect and prepare the information on the cost estimating procedures.

_____ Assemble the costs, routing sheets and estimating information into a sequentially filed set for each team.

_____ Deliver the completed package(s) to the Workshop Coordinator.

Figure B–4. Cost and Technical Data Preparation Checklist

_____ Selection of date

_____ Selection of Users/Customers

_____ Reserve a location for session

ARRANGE FOR:

_____ Rolls/Coffee/Soft-drinks

_____ Flip Chart or Blackboard

_____ Overhead Transparency Projector with a spare bulb. Must be _TRANSMISSIVE_, not _REFLECTIVE_ type. (Bulb in base, not in head.)

_____ Projection Screen, 5 X 5

_____ _Large_ conference table with chairs to seat members of the panel facing each other

_____ Place cards for each participant, visible from both sides

_____ Samples of the specific model of the Product or Process under study with accompanying literature

_____ Samples of competitive Products or Processes (where practical) and their sales literature

Figure B–5. _Preparation Checklist—Focus Panel_

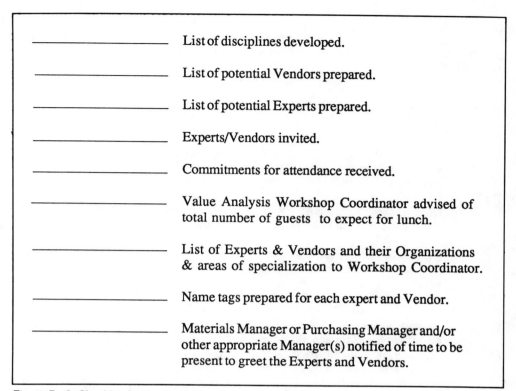

_____ List of disciplines developed.

_____ List of potential Vendors prepared.

_____ List of potential Experts prepared.

_____ Experts/Vendors invited.

_____ Commitments for attendance received.

_____ Value Analysis Workshop Coordinator advised of total number of guests to expect for lunch.

_____ List of Experts & Vendors and their Organizations & areas of specialization to Workshop Coordinator.

_____ Name tags prepared for each expert and Vendor.

_____ Materials Manager or Purchasing Manager and/or other appropriate Manager(s) notified of time to be present to greet the Experts and Vendors.

Figure B–6. *Checklist for Expert/Vendor Consultation*

Opening remarks should be straightforward and off-the-cuff (no notes) and should convey your honest feelings. We suggest that most of the following points be made:

1.   Tell them what project they will be studying.

2.   Describe the target goal for the study, in terms of Product Improvement *and* Cost Reduction.

3.   Tell them why they must succeed in this study. (This rationalization of the goal is essential to assure that the team members assume "ownership" of the goal. They must not be permitted to view the goal as "just another management stretch goal".)

4.   Indicate to the participants that you expect them to develop sufficient competence in the use of Value Analysis that they will use it on future problems.

5.   Describe the process used to select the participants. Emphasize that they were not merely "pulled from a list," but are regarded as "Key People" with specific capabilities which will contribute to solving the specific problems of the product under study. They are the "Movers & Shakers" of your organization.

6.   There are no constraints on what they may consider. This is their opportunity to suggest the investigation of changes which may have been suppressed in the past for fear that "Top Management wouldn't like it"." Emphasize that you expect them, at the Final Presentation, to recommend anything they think would benefit the organization. Tell them to leave the decision on controversial items to the Top Executives in attendance.

7.   Full-time attendance is required. There will be considerable homework between certain sessions.

Figure B–7. Guide to Keynote by Top Executive

WRITTEN BY: Top Executive

KEY POINTS:

A.   Comment on the past and/or present success of the organization and its people.

B.   The Value Analysis of this Product/Process is essential to the future of this organization. (Give reasons.)

C.   The form of Modern Value Analysis being used is primarily focused on User Needs. Cost reduction is a secondary, though essential, objective.

D.   List Workshop LOCATION, STARTING DATE, STARTING TIME.   Enclose the detailed Agenda. Emphasize that 100% participation is required.

E.   Emphasize the KEY or DECISION-MAKING status of all participants.

Figure B–8. Letter to Team Members

2.   LETTER INVITING VENDORS or EXPERTS TO THE "SYNTHESIS" SESSION

NOTE: Letter to Vendors is in normal type. For letter
                to Experts, substitute **boldface, [bracketed]** items.

WRITTEN BY: Materials or Purchasing Manager
                [Top Executive or other appropriate Executive]

KEY POINTS:

A.   "Thank you for agreeing to participate in our Value Analysis Workshop on _____".

B.   Several teams of our Key People (specify the numbers) from different areas of our organization (specify the areas) will spend a total of 100 hours over a period of two months in a fundamental analysis of our (name the Product or Process).

C.   The form of Value Analysis we are using emphasizes that we first make certain that the Product or Process fulfills the user's requirements, and then reduce the cost.

D.   You are one of a carefully selected group of vendors whom we have invited to help us in this effort by acting as critical sounding boards for ideas and concepts which the team members have agreed to "champion".

[D.   **You were carefully selected to help us in this effort by acting as a critical sounding board for ideas and concepts which the team members have agreed to "Champion"]**

E.   Your representatives should be prepared to give "on-the-spot cost estimates as well as technical advice as we discusssed. There will be space available for a small display of your products and descriptive literature.

[E.   **Delete ¶E for letter to Experts]**

F.   In addition, I would personally appreciate your making further suggestions, based on your broad experience and the application of your products in other areas.

[F.   **In addition, I would personally appreciate your making further suggestions, based on your broad experience.]**

G.   The Schedule:

>   Arrive 11:30 AM for a 15-minute briefing by the Value Analysis Coordinator.

>   Join the teams for lunch and then a free-interchange session from 1:00 PM to 4:00 PM.

H.   For your reference, I enclose data on the (Product/Process under study).

Figure B–9. Letter Inviting Experts to Synthesis Session

APPENDIX C: INDEX OF CASE STUDIES

The table on pages 289–295 references the salient points of the case studies found in chapter 13, and should prove useful for readers searching for data pertinent to their environments.

Case Study Number	Company or Organization and Project	Lesson Learned and Key Points
1	General Corporation heat pump. Check valve reduction.	Wild card works. An unconstrained list of words displayed on the creativity phase flip chart has a valuable triggering effect.
2	General Corporation heat pump. The 90-degree concept.	Unconstrained creativity pays off. The team environment of free interchange of information in the synthesis phase released the inhibitions of an advanced manufacturing engineer, who expressed his long-held opinion that angles other than 90 degrees cost money.
3	General Corporation heat pump. Enhance fins.	Old but good ideas never die. The power of the Idea Bank in capturing team thoughts during the function-cost allocation session is demonstrated in an item that the Project Engineer recorded on fin enhancements.
4	General Corporation heat pump. Keyholed and flanged side panels.	The value analysis target can be based solely on worth. The focus panel of users/customers identified a clear value analysis target: "enhance ac-

Case Study Number	Company or Organization and Project	Lesson Learned and Key Points
		cess." The creativity phase resulted in a major improvement in cost and in user acceptance.
5	General Corporation heat pump. Delete fan motor bracket clamp.	Experts/vendors have good ideas—use them. A motor vendor participating in the sounding board portion of the synthesis phase agreed to modify his product at no cost, eliminating a fabricated clamp.
6	General Corporation heat pump. Low-voltage block relocation.	Improving cost and worth concurrently. The user focus panel identified "reduce downtime" as a critical need. The cost to perform this function was low, indicating a value analysis target. This change improved customer satisfaction while further reducing cost.
7	General Corporation heat pump. Field conversion, side to down outlet.	Life-cycle cost advantage drives the decision to implement. Competition supplied a heat pump that was easily field-convertible from side to down. The user focus panel verified a strong customer desire for the feature. The value analysis system thus served to verify and implement a preconceived solution.
8	Engine manufacturer. 1,200 HP engine. Use low-cost spark plugs.	Value analysis targets trigger insight. Two sets of spark plugs were shipped with each engine. Both comprised $9.36 high-reliability plugs, for a total of $149.76. The team changed the set used only for run-in and mating tests to standard 89-cent plugs. The idea trigger was function-cost.
9	Engine manufacturer. 1,200 HP engine. Buy pushrod adjusting screws outside.	Make/buy. A very special hardened adjusting screw was tooled for in-house manufacture many years before. Its cost was reduced by 81 percent at a specialty screw-machine house. Again, function cost triggered a search by the manufacturing engineering manager.
10	Industrial oven manufacturer. Continuous-flow oven. Fiberglass doors.	You can always make a good product better. Focus panel participants identified heavy doors as low on their scale of likes. Lighter fiberglass doors are colored cast-iron gray.
11	Machine tool manufacturer. Milling machine. Cast-in lube oil header.	Improve value by changing outmoded company policy. An in-house foundry that was not equipped to perform no-

Case Study Number	Company or Organization and Project	Lesson Learned and Key Points
		bake molding had constrained this manufacturer of a horizontal milling machine from casting the oil header gallery as an integral part of the machine bed. The company went to an outside foundry. The redesigned bed saved 77 percent.
12	Engine manufacturer. 1,200 HP engine. Bearing temperature system.	Increased uptime improves value. The trigger for this was a user focus panel dislike: "oil leaks." This directed the team's attention to the piping in the bearing overtemperature system. The change eliminated all piping and, incidentally, saved $5,700.
13	Engine manufacturer. 1,200 HP engine. Unshielded ignition system.	Eliminate unwanted features and reduce cost. The user focus panel revealed that the standard shielded ignition system was not adding to the company's product acceptance. It was changed to an optional feature, saving $700 per engine.
14	Engine manufacturer. 1,200 HP engine. Water-cooled exhaust manifold.	Make it safe; improve value. Users rated hot manifolds (hard to work around) as a serious fault. The team found that changing to a water-cooled manifold reduced the cost by $5,000 per engine, due to a change from stainless steel to carbon steel.
15	Railway equipment supplier. Bus door actuator. Complete redesign.	Eliminate faults, give customers what they want, and improve price and market share. A discussion of the story told by Thomas F. Cook of his realization that knowing user-attitudes is essential to a valid modern value analysis effort.
16	Imodco Corp. CALM ocean buoy. Complete redesign.	Wild card works. In a classic example of the effectiveness of unconstrained function-focused creativity, the team doubled the number of watertight bulkheads, reducing the cost by $17,000/unit.
17	Mine equipment manufacturer. Drum coal miner. Cam-operated core breaker.	Function-cost analysis focuses the value problem. The company had developed a method of milling out a coal seam without leaving a core between the drums. It was inordinately proud of its accomplishment. The team discovered a $154,000 penalty

Case Study Number	Company or Organization and Project	Lesson Learned and Key Points
		for this pride. Focus panel data proved that users didn't really care. The team developed a patentable oscillating core cutter.
18	Titus Division, Philips Industries. Ceiling air diffuser. Complete redesign.	Informal user data works. With only the team members' subjective opinions of the needs of users/customers, this highly effective value analysis team reduced the diffuser's cost by $500,000 per year and significantly improved its performance.
19	Signode Industries. SPIRIT® Machine. Development of a new machine.	Value analysis optimizes a fundamental redesign. This powerhouse, four-team value analysis effort used the user focus panel coupled with a status arbitration session to develop a new generation of strapping machine, which has become the standard of the trade.
20	Magnaflux Corp. Testing Machine. Standardized module.	Function-cost focuses the problem. The value analysis team, in its analysis of cost and function, became convinced that lack of standardization was a major cause of nonfunctional cost in all of the company's products. The new standard tank frame saved $245 on the machine under study and $4,165 per year in direct spin-off to other machines.
21	Hobart Bros. Co. Welding rod. Change chemical formula.	Even a simple welding rod can benefit from value analysis. This leader in the welder and welding supplies industry undertook to value-analyze a simple welding rod, little more than a stick of steel with a chemical coating. Triggered by ten value analysis targets that were identified in the analysis phase, the team, which included the project engineer, proposed changes to reduce the cost per rod by 2 cents— for a $46,000 annual saving.
22	Automobile accessories manufacturer. Capital tooling. Restructure.	Value Analysis works on capital expenditures. Three value analysis teams were led by the three manufacturing engineers who had designed and estimated an $89 million manufacturing system to build 8,850,000

Case Study Number	Company or Organization and Project	Lesson Learned and Key Points
		controlled-displacement freon compressors annually. The effort resulted in a reduction in the capital requirement by over $28 million and a bonus reduction in unit cost of $3.69.
23	RV accessories manufacturer. Trailer dropstand. New product.	User data triggers a new product. The user focus panel revealed to the value analysis team that a significant need existed for a new kind of RV quick-deployment jack. The project engineer championed the new design.
24	Indiana Gas Company. Billing procedure. Minimize mailings.	Value analysis works on procedures. The value analysis team, in its analysis of this procedure, identified "render bill" as a value analysis target. Its creativity session focused on that function and resulted in a restructuring of the billing procedure, saving $72,000/year.
25	Commercial kitchen equipment manufacturer. Restaurant range. Stable stove spider.	User data triggers solution. A multi-state value analysis team achieved a cost saving of only 16 cents per range, but in the process eliminated a significant user-perceived fault. The new grate is both stable and rugged.
26	Valve manufacturer. Hydraulic valve. Foam pattern casting.	Vendor input pays off. During the sounding board session with vendors and experts, the process of foam pattern casting was discussed. This nearly net-shape process was new to the company and saved $65,000 per year.
27	Construction equipment manufacturer. High-travel forklift. Drive-through rear axle.	User data and function definition lead to a creative solution. The value analysis team accomplished a $952 saving per unit while eliminating a major reliability problem by replacing a long-drop transmission with a short-drop version. The trigger was the user focus panel fault data.
28	Butler Construction Company. Custom order flow. Project organization chart.	Improved communications improves value. The construction division of this major corporation assigned five key people to value-analyze this paperwork process. Two of the responses in the user focus panel directed them toward a restructuring of the authority chart on each project. Their new structure saves $1,110 per project.

Case Study Number	Company or Organization and Project	Lesson Learned and Key Points
29	Mid-West Conveyor. Post office conveyor system. Bolted hanger.	Construction project benefits from value analysis. Management charged the value analysis team with achieving four goals on cost reduction. Productivity, and in-sourcing. Essentially, all goals were reached or exceeded. An excellent example of the many proposals for cost reduction is the design of a new method of hanging to eliminate field welding.
30	Automobile parts manufacturer. Overhead study. New SPC limits.	Analysis of chart of accounts identifies value analysis targets. Two teams of key managers value-analyzed the overhead and SG&A expenses of their division of an international corporation. They implemented over 14 percent savings in the total of these costs! Many of their proposals existed before the VA effort, but one created during the team effort saved $65,000 per year through the rearrangement of SPC limits.
31	Butler Manufacturing Company. Insl-Wall® system. Substitute brads.	Cost reduction improves appearance. The value analysis team made use of the user focus panel to establish an unfulfilled user desire: the appearance of the panel. The team replaced all of the staples and T-nails with brads, improving its appearance and saving over $3,000 per year.
32	Butler Manufacturing Company. Insl-Wall® system. Shiplap panel.	User need to "facilitate erection" of the panels results in cost reduction and improved acceptance. The same team described in case study 31 found that the user focus panel data also indicated a strong user desire to minimize erection labor. By concentrating on the value analysis target "facilitate erection," the team developed a shiplap panel joint that greatly eased the erection process and presented the company with a bonus $25,000-per-year saving.
33	Butler Buildings. Delta joist®. Reinforcing sleeve.	Broadened viewpoint of function-cost and function worth leads the way to product improvement. The value analysis team identified a value analysis target by relating function-cost to function-worth data developed in a user focus panel. One of the proposals built upon the target "increase

Case Study Number	Company or Organization and Project	Lesson Learned and Key Points
		strength" by developing a sleeved reinforcement system as standard for all plants.
34	Walker Manufacturing Company. Trenchduct®. Shield adjustment.	User viewpoint guides team to a combined cost reduction and product improvement. The five key employees assigned to value-analyze this major company product identified "establish screed" as a value analysis target. Its cost was 11.8 percent of the total and the user focus panel data revealed the function as important. The redesign eliminated all of the screwdriver work in adjusting the U compartment to retain poured concrete.
35	Commercial kitchen equipment manufacturer. Restaurant range. Oven door restyling.	Where style controls, value analysis delivers style at minimum cost. The same team that accomplished the stable grate of case study 25 deigned to attack a sacred cow and succeeded in saving $51,000 over a three-year period while maintaining the user-perceived value of a sculpted oven door.
36	Security systems manufacturer. Electronic circuit. Complete redesign.	Value analysis works on electronic circuitry. Function-cost allocation drew the attention of the team to MTBF and the cost of the drive circuits. A new approach to power switching and the creation of a diagnostic display function on the board saved nearly $21,000 the first year and improved user acceptance.

REFERENCES

Alexander, Tom. 1965. "Synectics: Inventing by the Madness Method," *Fortune* (August).

Borden, Richard C. 1935. *Public Speaking as Listeners Like It*, New York: Harper Bros.

Crosby, Philip B. 1979. *Quality Is Free*, New York: McGraw-Hill.

Dixon, Robert L. 1953. "CREEP," *The Journal of Accountancy* (July).

Drucker, Peter. 1979. *Adventures of a Bystander*, New York: Harper & Row.

Fallon, Carlos. 1971. *Value Analysis to Improve Productivity*. New York: Wiley Interscience.

Koestler, Arthur. 1968. *The Act of Creation*. New York: MacMillan.

Levitt, Theodore. 1963. "Creativity Is Not Enough," *Harvard Business Review* (May–June).

Miles, Lawrence D. 1961. *Techniques of Value Analysis and Engineering*, New York: McGraw-Hill.

Miles, Lawrence D. 1972. *Techniques of Value Analysis and Engineering, 2nd Edition*, New York: McGraw-Hill.

Osborn, Alex F. 1953. *Applied Imagination*, New York: Charles Scribner's Sons.

Peters, Tom, and Nancy Austin. 1985. *A Passion for Excellence*, New York: Random House.

Stevenson, B. 1967. *The Home Book of Quotations*, New York: Dodd Mead & Co.

INDEX

Page references in *italics* refer to figures; page references in **bold** refer to tables.